T0228939

ADVANCES IN
DNA SEQUENCE SPECIFIC AGENTS

Volume 2 • 1996

ADVANCES IN
DNA SEQUENCE SPECIFIC AGENTS

Series Editor: LAURENCE H. HURLEY
The Drug Dynamics Institute
The University of Texas at Austin

Volume Editor: JONATHAN B. CHAIRES
Department of Biochemistry
University of Mississippi Medical Center
Jackson, Mississippi

VOLUME 2 • 1996

 JAI PRESS INC.

Greenwich, Connecticut   London, England

CONTENTS

LIST OF CONTRIBUTORS

Karen Alessi

Department of Chemistry
New York University
New York, New York

Jonathan B. Chaires

Department of Biochemistry
University of Mississippi Medical Center
Jackson, Mississippi

Nicholas Farrell

Department of Chemistry
Virginia Commonwealth University
Richmond, Virginia

David E. Graves

University of Mississippi
University, Mississippi

Paul B. Hopkins

Department of Chemistry
University of Washington
Seattle, Washington

Luis A. Marky

Department of Chemistry
New York University
New York, New York

Stephen Neidle

The CRC Biomolecular Structure Unit
The Institute of Cancer Research
Sutton, Surrey, England

Don R. Phillips

School of Biochemistry
La Trobe University
Bundoora, Victoria, Australia

Dionisios Rentzeperis

Department of Chemistry
New York University
New York, New York

John O. Trent   The CRC Biomolecular Structure Unit
                                         The Institute of Cancer Research
                                         Sutton, Surrey, England

Andrew H.–J. Wang   Biophysics Division
                                         Department of Cell and Structural
                                         Biology
                                         University of Illinois at
                                         Urbana-Champaign
                                         Urbana, Illinois

PREFACE

At the interface of chemistry and biology it is difficult to imagine a subject of more importance than molecular recognition. It is therefore surprising that this is a relatively new discipline, although the 1987 Nobel Prize in Chemistry was awarded to Donald Cram and Jean-Marie Lehn for their pioneering of this area. While chemists have designed and synthesized quite elaborate models to study host–guest chemistry, it is obvious that we have a long way to go before we can approach the specificity of Nature—not surprising, since Nature has been playing this game since the very start of life.

DNA sequence specificity is a sub-specialty in the general area of molecular recognition. This area includes macromolecular–molecular interactions (e.g., protein–DNA), oligomer–DNA interactions (e.g., triple strands), and ligand–DNA interactions (e.g., drug–DNA). It is this latter group of DNA sequence specificity interactions that is the subject of Volumes 1 and 2 of *Advances in DNA Sequence Specific Agents*. As was the case for Volume 1, Part A also covers methodology, but in Volume 2 we include calorimetric titrations, molecular modeling, X-ray crystallographic and NMR structural studies, and transcriptional assays. Part B also follows the same format as Volume 1 and describes the sequence specificities and covalent and noncovalent interactions of small ligands with DNA.

This volume is aimed in general at scientists who have an interest in deciphering the molecular mechanisms for sequence recognition of DNA. The methods have general applicability to small molecules as well as oligomers and proteins, while the examples provide general principles involved in sequence recognition.

We would like to acknowledge the efforts of Mr. David Bishop, who shares in the responsibilities of proofreading and editing the manuscripts for the individual chapters, and whose assistance was invaluable in assembling the final materials for publication.

Laurence H. Hurley
Series Editor

Jonathan B. Chaires
Volume Editor

PART I

METHODS USED TO EVALUATE THE
MOLECULAR BASIS FOR SEQUENCE
SPECIFICITY

CALORIMETRIC STUDIES OF
DRUG–DNA INTERACTIONS

Luis A. Marky, Karen Alessi, and

Dionisios Rentzeperis

Advances in DNA Sequence Specific Agents
Volume 2, pages 3–28.
Copyright © 1996 by JAI Press Inc.
All rights of reproduction in any form reserved.
ISBN: 1-55938-166-3

I. INTRODUCTION

The function of nucleic acids is carried out through interactions with other molecules. In the readout of the genetic information and in the control of gene regulation, the interaction of proteins and other ligands with DNA is essential. Distinctive conformations are associated with the function of DNA; sites of interaction of DNA with gene regulatory proteins and smaller ligands have been found to exhibit characteristic conformational properties.[1] For a complete understanding of how nucleic acids carry out their biological role, it is imperative to have detailed physical information on these interacting systems. For these reasons, there is now unprecedented interest in the conformational fine structure of nucleic acids and the forces that determine this structure. On the other hand, tight binding of drugs to DNA also underlies a number of important biological phenomena including mutagenesis, carcinogenesis, and antitumor activity of some therapeutic agents.[2] The complete understanding of these interactions includes a correlation between structure, thermodynamics, and kinetic data. The structure of a drug–DNA complex provides an implicit description of specific molecular interactions, whereas thermodynamics addresses the molecular forces that control the affinity and allows us to parse the relative contributions from base stacking, hydrogen bonding, electrostatic, and/or hydrophobic interactions. Kinetics probes the time course of formation or dissociation of a particular drug–DNA complex.

In the specific case of determination of the overall energetics of the binding of a particular ligand to a DNA molecule, the approach that has been used by investigators is to obtain *standard* thermodynamics profiles: ΔG^0, ΔH, and ΔS.[3–8] First is the determination of binding affinities, K_b, indicating how tight the formation of the particular ligand–macromolecule complex is and yielding the overall free energy value ($\Delta G = -RT \ln K_b$) that tells us how favorable the formation of the ligand–DNA complex is. This information alone normally allows the investigator to make chemical modifications on the ligand (or in the DNA) with the idea of producing a more specific ligand that is competent to form a tighter ligand–DNA complex. Second is the measurement of the interacting heat that dissects the free energy term into its component enthalpy and entropy contributions. The heat measurements can be obtained indirectly by the use of the van't Hoff equation from the dependence of K_b with temperature, or directly by use of calorimetric techniques; in this case, the entropy can be calculated from the Gibbs equation ($\Delta G = \Delta H - T\Delta S$). In a simplistic and more general view, the resulting enthalpy contributions tell us about the formation or disruption of chemical interactions, whereas the magnitude of the entropic term measures the degree of order or disorder of the system.

It should be emphasized that the *complete* thermodynamic characterization of the formation of a particular ligand–DNA complex, starting with the physical mixing of a ligand to a DNA molecule at selected experimental conditions, involves the relative energetic contribution to both heat and entropy from all the observed molecular interactions of the participating molecules. These include the formation or disruption of van der Waals and hydrogen bonds, electrostatic interactions, release or uptake of counterions, and the release or uptake of water molecules. In addition, it is important to take into account both the actual molecularity of the binding reaction, which may be related to the stoichiometry of the complex, and the changes in the conformation of the macromolecule, if any, upon complexation with a ligand.

In this report, we discuss the thermodynamic interpretation of physical binding data of ligands with different binding modes and their sequence specificity, and ligand binding to monomolecular single and double hairpins, as well as to multistranded DNA structures. The instrumentation used for the measurement of the interacting heats is also mentioned, and the several systems discussed here will be those of calorimetric investigations, mostly from the authors' own work on interactions of the intercalator ethidium bromide with DNA and the minor groove ligand netropsin (see Figure 1). These ligands were chosen for the following reasons: (1) ethidium and netropsin have been studied extensively by a wide variety of physical techniques,[5,8-26] (2) their DNA complexes illustrate two well-known types of binding modes that are stabilized by a mixture of molecular interactions, and (3) these ligands have proven to be ideal in calorimetric measurements because they do not aggregate at the concentrations used, thereby eliminating a large heat contribution from self-association. In addition, ethidium is widely used

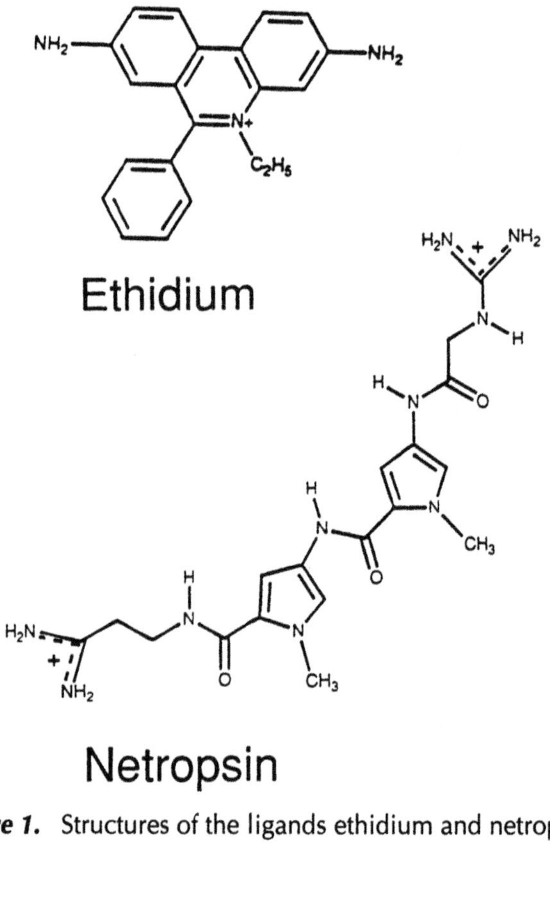

Figure 1. Structures of the ligands ethidium and netropsin.

in molecular biology as a stain for DNA because of its increased fluorescence quantum yield when bound to duplex DNA.[27]

II. HIGH-SENSITIVITY CALORIMETRIC TECHNIQUES

Several types of calorimetric techniques have been employed to measure directly the heat of interaction of ligands to DNA molecules—these include isothermal titration calorimetry (ITC),[16,28–30] batch calorimetry,[4,5] and stopped-flow calorimetry,[6] which are all carried out at constant temperature—and indirectly by differential scanning calorimetry (DSC).[5] Because these techniques have been described in detail elsewhere,[31–33] in this section the ITC and DSC techniques that the authors' group uses routinely will be described briefly. Not only are these techniques the most effective ones in the determination of the interacting heats, but they complement each other in the complete understanding of the energetics of the particular system under study.[29,30,34,35]

A. Isothermal Titration Calorimetry (ITC)

We use a Microcal Omega instrument to measure the heat evolved as a function of the amount of titrant for the association of ligand to a macromolecule[33]; a typical titration is shown in Figure 2. Two mL (the effective volume is actually ~1.4 mL) of a DNA solution is placed in the reaction cell and titrated with a 20–40-fold higher concentration of the ligand solution that is placed in a 50- or 100-μL syringe. Complete mixing is effected by stirring of the syringe paddle at 400 rpm. The heat is measured differentially against a reference cell that is filled with water or buffer. The instrument is initially calibrated by means of known standard electrical pulses. Normally, an average of 15–20 injections, 4–10 μL each, is needed in a single titration experiment; the precision of the resulting heat of each injection is less than 0.5 μcal. The binding enthalpy, ΔH_b, can be determined in two ways: (1) by averaging of the areas of the resulting initial peaks pertaining to a particular site that follow each injection and correction for the dilution heat of the ligand and

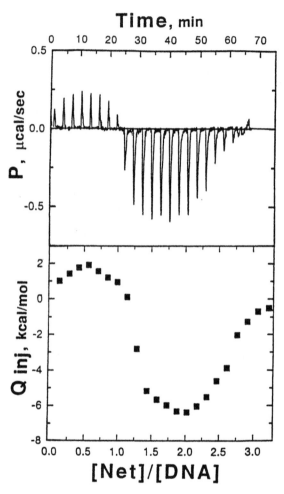

Figure 2. Typical calorimetric titration at 7 °C. Top panel shows the resulting experimental curve for the titration of 1.4 mL of a 36 μM (in total strands) solution of [d(GGA₅T₅CC)]₂ with 7-μL injections of 0.9 mM solution of netropsin. The corresponding heats per injection as a function of the netropsin/duplex molar ratio are shown in the lower panel. These results clearly show the presence of multiple sites and that the binding enthalpy to the first site is endothermic, whereas binding to the second site is exothermic. This is consistent with the endothermic contribution of the removal of electrostricted water molecules from the minor groove of the oligomer duplex.

normalization by the concentration of added titrant, and (2) by analysis of the calorimetric binding isotherm from which, in most cases, one is able to obtain, in addition to the binding enthalpy, binding affinities and the overall stoichiometry of the complexes. The experimental calorimetric binding isotherm is the dependence of the total heat, Q_T, on the total concentration of ligand added, X_T. These experimental data, or dQ_T/dX_T vs. X_T, can be fitted by defining a thermodynamic binding model for the reaction of interest involving single or multiple sites—each type of site can be defined by three parameters: K_b, ΔH_b, and n (number of ligands per site). The above three parameters for each type of site may be determined iteratively by use of the Marquardt algorithms that are incorporated in the software of this calorimeter. The fitting functions for one and two independent sites have been described previously.[33,36] The best initial fitting procedure is to fix the enthalpy value obtained from the initial injections and/or the value of n, which can be determined using spectroscopic methods, or to let all three parameters float until the lowest standard deviation of the fit is obtained.

B. Differential Scanning Calorimetry (DSC)

This technique measures the heat capacity of a sample solution relative to that of the same buffer as a function of temperature, usually ranging from 0 to 100 °C. After subtraction of a buffer vs. buffer scan from that of the sample vs. buffer and normalization to the number of moles of solute, the area of the resulting curve (equal to $\int \Delta C_p dT$) is proportional to the enthalpy (ΔH_{cal}) of the order–disorder transition of a DNA helical sample, for example. Conversion of the experimental curve to that of a $\Delta C_p/T$ vs. T curve and evaluation of its area (equal to $\int \Delta C_p/T dT$) allows for the determination of the transition entropy. Thus, one is able to obtain the overall stability of the sample, T_m, and a standard thermodynamic profile: ΔG^0, ΔH, ΔS, and the ΔC_p between initial and final states. In addition, from analysis of the shape of a DSC curve, one is able to test if the unfolding reaction for oligonucleotide structures takes place in a two-state

transition or through the formation of intermediate states by
simple comparison of the calorimetric enthalpy with that of the
model-dependent or van't Hoff enthalpy, ΔH_{vH}. This enthalpy
is best determined from the temperatures that correspond to the
half-height of the experimental curve: if the $\Delta H_{vH}:\Delta H_{cal}$ ratio is
equal to 1, then the transition takes place in a two-state manner;
if it is less than 1, the reaction takes place through intermediate
states.[31,32,37] For polynucleotides, the resulting $\Delta H_{vH}/\Delta H_{cal}$ value
corresponds roughly to the size of the cooperative unit. The
authors recommend that DSC scans be obtained for nucleic acid
samples used in ligand studies as a routine check for determination of sample purity, stability, and helix-coil enthalpies. The
concentration needed for carrying out these experiments is
small, about 0.5 mg for polynucleotides and 2 mg for oligonucleotides.

Furthermore, the thermodynamic release of counterions,
Δn_{Na+}, for the order–disorder transition of a particular nucleic
acid structure can be determined directly from DSC scans at
different salt concentrations. The relevant equation[38] is
$\Delta n_{Na+} = (\Delta H/RT_m^2) \, dT_m/d\ln(Na^+)$, where all terms are determined
either directly from DSC experiments or in combination with
optical melting curves, as has been done extensively by the
authors' group.[28,29,34,35,37]

Another application of the DSC technique is that of obtaining
standard thermodynamic profiles for the association of ligands.
This experimental protocol consists of a scan of a fully saturated
ligand–DNA sample vs. a DNA sample with identical concentration in the reference cell, and a typical curve for the complexation of ethidium to poly(rA)·poly(rU) is shown in Figure
3. Since the preferential binding of a ligand to the helical state
normally shifts the helix-coil transition of the DNA to higher
temperatures, one first obtains a negative peak corresponding
to the transition of the unligated DNA and a second positive
peak at higher temperatures that corresponds to the unfolding
of the ligand–DNA complex. The complete separation of these
two peaks depends on the binding affinity of the ligand. If special care has been exercised in both the dialysis of the DNA
sample and the matching of the DNA concentration in the two

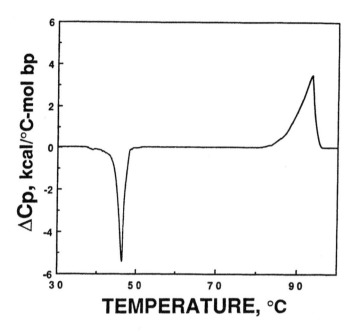

TEMPERATURE, °C

Figure 3. Typical DSC scan of a poly(rA)·poly(rU) solution saturated with ethidium (4:1, in Pi per ethidium) vs. a poly(rA)·poly(rU) solution; the buffer consisted of 10 mM NaPi at pH 7. The concentration of polynucleotide in both cells is identical and equal to 3.2 mM (in phosphate). Analysis of this curve yields ΔH_{cal} = 6.9 kcal mol^{-1} (in bp) and ΔH_{vH} = 720 kcal mol^{-1} for the free macromolecule, ΔH_{cal} = 11.9 kcal mol^{-1} (in bp), and ΔH_{vH} = 260 kcal mol^{-1} for the ethidium–polynucleotide complex. The difference in heats corresponds to an ethidium dissociation enthalpy of 10 kcal mol^{-1}. The increase in thermal stability by about 52 °C of the ethidium–poly(rA)·poly(rU) complex relative to the free duplex yields a K_b of 2.2 × 10^5 M^{-1} when extrapolated with the van't Hoff equation to 25 °C.

cells, the initial baseline should correspond exactly to the natural baseline of the instrument, preventing a shift of the sample baseline (vs. a buffer) to a lower position due to the known exclusion of solvent molecules by macromolecules. Integration of the areas of each peak yields the associated enthalpies for the unfolding of the free DNA and ligand–DNA complex, and the enthalpy difference corresponds to the enthalpy of dissociating the ligand from the complex. This approach uses a thermodynamic cycle in which the contribution from the single strand cancels out. The determination of this dissociating enthalpy has a high uncertainty, of at least 15%, because it corresponds to the difference of two large numbers. The temperature difference of the two peaks, ΔT_m, is proportional to the difference in binding affinities of the ligand toward the helical, K_h, and coil states, K_c, of the DNA and follows the equation

$$\Delta T_m = (T_m^0 T_m / R\Delta H) [B_h \ln(1 + K_h a_L) - B_c \ln(1 + K_c a_L)]$$

where T_m^0 and T_m are the transition temperatures of the free and saturated complex, respectively; B_h and B_c are the number of bound ligands (per base pair) to the helical and coil states, respectively[39]; R is the universal gas constant; ΔH is the transition enthalpy of the unligated DNA; K_h and K_c are the ligand binding affinities to the helical and coil states, respectively; and a_L is the activity of the free ligand at $T = T_m$, which is normally assumed equal to one-half of the total concentration of ligand. An additional assumption is that binding to the coil states is negligible because these coil states are present at high temperatures, so the terms that are involved in the coil states can be neglected. These ligand binding affinities may be extrapolated to the temperature of interest by the van't Hoff equation $d\ln K_b / dT = \Delta H_b(T) / RT^2$, where $\Delta H_b(T)$ is the binding enthalpy at T, equal to $\Delta H_b(T_m) + \int \Delta C_p dT$. Alternatively, one can assume $\Delta C_p = 0$ and use the binding enthalpy from ITC experiments. One note of caution when using this approach: it is best to use a helix-coil ΔH value equal to the unfolding of the base pairs that are actually covered by the bound ligand; otherwise, the resulting K_b values are inflated. One type of example that may illustrate this effect is the specific binding of actinomycin to the set of DNA oligomer duplexes d[(AT)$_n$GC(AT)$_n$]$_2$ (or netropsin to d[(GC)$_n$AATT(GC)$_n$]$_2$). Because each ligand binds tightly or covers the four central base pairs of the oligomers, a change of n from 1, 2, and 3 would leave ligand-free dA·dT base pairs (or dG·dC in the case of netropsin). The K_b values should be similar for all three oligomers. Also, it is best, when possible, to use a ΔH_b determined by titration calorimetry at several temperatures because heat capacity effects can be included.

III. INFERENCE OF BINDING MODE FROM THERMODYNAMIC PROFILES

A great variety of useful antitumor drugs act by binding directly to double-helical DNA, interfering with both replication and transcription. Some drugs intercalate between adjacent base pairs (ethidium, daunomycin), others bind within the minor groove of B-DNA (netropsin, distamycin), and a third class consists of those that interact nonspecifically with DNA. The detailed structural properties of drug–DNA complexes revealed by both crystallographic and NMR methods have shown how each drug binds to DNA and its specific molecular interactions in the complex. However, such structural pictures, in general, tell us nothing about (1) the nature of the overall molecular forces that drive complex formation in solution, (2) the relative energetic contributions of the specific molecular interactions, (3) the behavior of these interactions with temperature and pressure changes, and (4) hydration effects. To complement these structural studies, it is necessary to carry out thermodynamic as well as kinetic measurements. The ultimate goal of these investigations is the design of better drugs, which can be deduced if we have a complete thermodynamic library of drug–DNA interactions. This particular library should include drugs with different binding modes and different binding affinities and specificities as well as the relative contribution of specific chemical groups.

What can we learn from the thermodynamic profile of a particular ligand–DNA system? A quick answer to this question is that a lot can be learned; for instance, Table 1 presents several examples of standard thermodynamic profiles for the binding of drugs, with four different binding modes, to two synthetic polydeoxynucleotides, poly-[d(AT)]·poly[d(AT)] and poly[d(GC)]·poly[d(GC)], that are used in a simple test

Table 1. Thermodynamic Profiles for Ligand Binding to Synthetic
Polydeoxynucleotides with Different Binding Modes

Polymer	K_b (M^{-1})	ΔG_b^0 $(kcal\ mol^{-1})$	ΔH_b^0 $(kcal\ mol^{-1})$,	$T\Delta S_b$ $(kcal\ mol^{-1})$	$dln\ K_b/$ $dln\ [Na^+]$
Ethidium (intercalation)[a]					
d(AT)·d(AT)	4.6×10^6	−9.1	−10.0	−0.9	−1.1
d(GC)·d(GC)	2.0×10^6	−8.6	−6.3	+2.3	−0.9
Dipyrandium (partial insertion)					
d(AT)·d(AT)[b]	5.4×10^4	−6.5	+4.2	+10.7	−1.1
d(GC)·d(GC)	1.0×10^4	−5.7	+4.8	+10.5	−1.0
Dipyrandenium (outside)					
d(AT)·d(AT)[c]	3.4×10^4	−6.2	+3.4	+10.3	−1.1
d(GC)·d(GC)	1.0×10^4	−5.5	+0.7	+6.2	−0.9
Netropsin (minor groove)					
d(AT)·d(AT)[d]	5.8×10^9	−13.1	−12.2	+0.9	−1.6
d(GC)·d(GC)	1.5×10^5	−6.9	−4.3	+2.6	−1.5
Distamycin (minor groove)[e]					
d(AT)·d(AT)	3.5×10^9	−12.8	−19.1	−6.3	−1.1
d(GC)·d(GC)	3.3×10^5	−7.4	−5.1	+2.3	−0.9

Notes: [a]Values were obtained from a combination of spectroscopic and/or ITC experiments in 10 mM sodium phosphate buffer, 1 mM Na2EDTA, pH = 7.0: [a]from Ref. 3; [b]from Ref. 7; [c]from Ref. 4; [d]from Ref. 16; [e]from Ref. 28; all other values were determined in the authors' laboratory (unpublished results).

of ligand specificity for dA·dT or dG·dC base pairs. In these four fundamental types of binding modes, carried out under similar solution conditions, the binding affinities varied with the binding mode: about 10^6 for intercalators, about 10^4 for outside and partial insertion binders, and ranging from 10^5 to 10^9 for minor groove binders. The class of minor groove binders is the only one that shows a strong specificity for dA·dT base pairs. The dissection of the free energy terms into their component enthalpic and entropic contributions reveals the overall nature of the driving forces. For intercalators and minor groove binders these forces are primarily enthalpic, whereas for outside binders and ligands that partially insert into DNA they are entropic.

Therefore, knowledge of the thermodynamic profile for a small ligand allows us to distinguish between ligands that do penetrate into the helix and those that do not. The magnitude of the enthalpy is also important to consider: a large and favorable term may indicate the exclusive formation of van der Waals interactions, whereas an unfavorable or small favorable enthalpy might include an extra contribution of removal of electrostricted water molecules.[40] This latter contribution can be obtained from measurements of the volume change (ΔV) associated with the reaction of a ligand to DNA. This may be determined by use of volumetric techniques such as dilatometry or magnetic suspension densimetry (a review[41] of the latter has been presented elsewhere). The amount of counterions released upon binding of a cationic ligand to a DNA molecule is also important to know because, in general, this term tells us about the effective number of charges on the ligand that are contributing to the overall stability of the ligand–DNA complex by electrostatic effects. If DNA oligomers are used in the binding studies, it is also important to take into account the overall charge density of the oligomer that can be estimated from the values of Δn_{Na^+} determined directly in DSC experiments.

IV. INTERACTION OF LIGANDS WITH SHORT DNA HAIRPINS

Hairpin structures are a common feature of RNA molecules,[42,43] and they have been postulated to form in DNA molecules in regions with palindromic sequences, which are implicated in gene regulation.[44–46] These hairpin molecules are useful for thermodynamic studies because they form partially paired duplexes that are stable over a convenient temperature range and melt in monomolecular transitions. Ethidium binds to DNA duplexes with a sequence preference for a 5'-Pyr-Pur-3' site and neighbor exclusion of at least two base pairs.[14] It is thus possible to design hairpin molecules with helical stems of four base pairs containing single binding sites and to study sequence specificity as well as the effect of loop size on the thermodynamics of ethidium binding to these molecules. In addition, the short length of their helical stems shows a reduced T_m-dependence on salt concentration that nearly eliminates the electrostatic contribution of these molecules toward the binding of cationic ligands.

A. Ethidium Specificity and Binding to Thymine Loops

Table 2 shows the thermodynamic profiles, measured at 10 °C, for the association of ethidium to a set of four hairpin molecules containing a loop of five thymines with the following 5' to 3' stem sequences: GTAC, CATG, GCGC, and CGCG. The results indicate that the stems bind one or two ligands following the 5'-Pyr-Pur-3' sequence preference, whereas the thymine loops accommodate two ligand molecules. Ethidium binding to these base-pair stacks is in the order of CG/CG > TA/TA ≈ CA/TG. Thus, at this low salt concentration, ethidium has a slight preference for the CG/CG base-pair stack. Relative to the stem sites, the binding affinity to the loops is lower by at least one order of magnitude. In all cases the association of ethidium to the stem or loop is primarily enthalpy driven, with a more favorable enthalpy contribution for those hairpins that bind only one ligand

Table 2. Thermodynamic Profiles for Ethidium Binding to Hairpin Loops[a]

Hairpin (Stem)	Site	n	K_b (M^{-1})	ΔG_b^0 (kcal mol^{-1})	ΔH_b (kcal mol^{-1})	$T\Delta S_b$ (kcal/mol)
GTAC	Stem	1.0	4.9×10^5	−7.4	−12.4	−5.0
	Loop	2.0	1.1×10^4	−5.2	−9.2	−4.0
CATG	Stem	2.0	9.1×10^5	−7.7	−9.1	−1.4
	Loop	2.0	4.3×10^4	−6.0	−7.8	−1.8
GCGC	Stem	1.0	3.9×10^6	−8.5	−10.1	−1.6
	Loop	2.0	6.3×10^4	−6.2	−8.6	−2.4
CGCG	Stem	2.0	3.2×10^6	−8.4	−8.1	+0.3
	Loop	2.0	5.1×10^4	−6.1	−7.6	−1.5

Notes: [a]All values were obtained from the fit of calorimetric titrations, the standard deviation of these fits ranging from 2 to 7%, in 10 mM sodium phosphate buffer, 0.1 mM Na₂EDTA, pH 7.0, 10 °C. The ΔG^0 values are within 5%; ranges for the other values are given in parentheses: ΔH^0 (±3%), and $T\Delta S^0$ (±8%).

Table 3.   Thermodynamic Profiles for Ethidium Binding to Hairpin Loops[a]

Hairpin	Site	n	K_b (M^{-1})	ΔG_b^0 $(kcal\ mol^{-1})$	ΔH_b $(kcal\ mol^{-1})$	$T\Delta S_b$ $(kcal\ mol^{-1})$	$d\ln K_b/$ $d\ln [Na^+]$
T3	Stem	1.0	1.2×10^6	−8.2	−8.6	−0.4	−0.32
	Loop	2.0	2.9×10^4	−6.0	−6.6	−0.6	
T5	Stem	0.9	1.7×10^6	−8.4	−9.8	−1.4	−0.53
	Loop	1.8	6.5×10^4	−6.5	−8.9	−2.4	
T7	Stem	1.0	2.5×10^6	−8.6	−11.6	−3.0	−0.99
	Loop	2.0	3.5×10^4	−6.1	−12.7	−6.6	

Note: [a]All values were obtained from the fit of calorimetric titrations in 10 mM sodium phosphate buffer, 0.1 mM Na₂EDTA, pH 7.0, 20 °C (see Ref. 29 for details).

molecule. This can be explained for the stem sites in terms of the neighbor exclusion parameter, i.e., two ligands will distort the helical stem to a greater extent. For the sites in the loops, the proximity of the thymines to the stem may favor the partial formation of an extra base-pair stack. Therefore, ethidium binding to the "CT/TG" base-pair stack on the loop side is more favorable than the association to the "GT/TC" or to the constrained thymines of the loops.

A better understanding of the binding of ethidium to these thymine loops has been shown by Rentzeperis and co-workers.[29] This analysis used a set of hairpins with sequence d(GCGCT$_n$GCGC), n = 3, 5, and 7, designated as T$_3$, T$_5$ and T$_7$ hairpins, respectively. The resulting thermodynamic profiles, from deconvolution of the titration binding isotherms, are shown in Table 3. Each hairpin contains two sets of binding sites that correspond to one ligand in the stem with binding affinity, K_b, of about 1.8×10^6 M^{-1}, and two ligands in the loops with K_b of about 4.3×10^4 M^{-1}. The similar binding affinities for the association of ethidium to the helical stem of these hairpin molecules correspond to the intercalation in the center CG/CG base pair stack of this helical stem,[14] whereas the lower K_b values for the loop sites correspond to nearly nonspecific binding of this ligand. The average ΔG_b^0 of −8.4 kcal mol^{-1} and −6.2 kcal

mol^{-1} for the stem and loop sites, respectively, correspond to binding processes that are primarily enthalpy driven. However, the binding enthalpy, ΔH_b, ranges from -8.6 (T_3) to -11.6 kcal mol^{-1} (T_7) for the stem site and -6.6 (T_3) to -12.7 kcal mol^{-1} (T_7) for the loop sites. This suggests that an increase in the length of the loop from three to seven thymine residues improves the overall stacking interactions of the ligand with the DNA base pairs, as seen by the increase in binding enthalpy of -1.2 kcal mol^{-1} and -3.0 kcal mol^{-1} for a step increase of two and four thymine residues, respectively. Alternatively, the local unwinding of the helical stem upon ethidium binding induces an increase in stacking interactions of the loop thymines. Therefore, the presence of the loops has a small influence on the local helical structure of the stem of these hairpins, which is consistent with the results of additional calorimetric binding experiments of ethidium to the T_5 hairpin, in the temperature range of 5 °C to 30 °C, that have shown a heat capacity effect of $\Delta C_p = -87$ cal K^{-1} mol^{-1} for the binding to the stem. An increase in the length of the loops results in a favorable increase in the binding enthalpies that is nearly compensated by unfavorable entropic terms. The small decrease in the favorable ΔG_b^0 terms with the increase in number of thymines in the T_5 and T_7 hairpins, relative to the T_3 hairpin, corresponds to a near enthalpy–entropy compensation for the two types of binding sites, equal to $\Delta \Delta H_b = \Delta (T\Delta S_b) = -0.75$ kcal mol^{-1} for the stem sites and $\Delta \Delta H_b = \Delta (T\Delta S_b) = -1.5$ kcal mol^{-1} for the loop sites. These differential compensating terms for the loop sites are twice the values of the stem sites and consistent with the higher ordering of ethidium molecules in the loops of these molecules. These thermodynamic results cannot predict the structure of the weaker sites. We speculate that the thymines in the loop of the T_3 hairpin are outside or completely crowded inside the loop; an increase in the number of residues to five and seven would result in a rearrangement of these thymines inside the loop with the formation of additional stacking interactions, which are favorable for ethidium binding. These additional sites may be composed of the C·G base pairs at the end of the helical stem plus the adjacent two thymines of the loops, forming a local duplex structure of a CT/TG base-pair stack and the constrained loop thymines.

B. Netropsin Binding to Double Hairpins Containing Nicks and Thymine Loops

Netropsin is a charged basic oligopeptide with a wide range of antibiotic activities against bacteria, fungi, and viruses,[47] but, because of its toxicity, it is not used in clinical studies. This ligand binds in the minor groove of double-helical B-DNA and shows a strong sequence specificity for dA·dT base pairs.[48–50] The molecular basis for the formation of netropsin–DNA complexes has been studied by a variety of techniques.[5,8,16–26] The strong specificity for four to five dA·dT base pairs is attributed both to the specific formation of hydrogen bonds between the amide protons of netropsin with N-3 of adenine and O-2 of thymine in the nucleotide edges facing the floor of this groove[22,26] and to van der Waals contacts between the methylpyrrole groups of netropsin and the sugar phosphate backbone of DNA resulting from its tight fit in the minor groove. The molecular interactions observed in structural studies have been correlated with the molecular forces obtained in thermodynamic studies; i.e., deep penetration in the minor groove of synthetic DNA molecules is accompanied by high binding affinities, exothermic enthalpies, or favorable binding entropies.[8]

In spite of the intensive investigations, important questions still remain unanswered concerning sequence-dependent conformational preferences because of deficiencies in the oligomeric and polymeric model systems. Some of these problems include the disproportionate effect of terminal ends on the physical properties under investigation and the fact that bimolecular melting transitions of most helices do not resemble the monomolecular melting within a natural nucleic acid polymer. Another limitation is that the melting of most oligomeric duplexes yields "linear" single strands since the ends are "open," whereas the local melting within a long nucleic acid molecule yields an interior "bubble." A double-hairpin-shaped molecule would thus solve most of these problems. Another important aspect of unligated double-hairpin structure is that it will demonstrate whether the phosphodiester gap significantly affects netropsin binding. In addition, our ultimate goal is to understand the stability and conformational plasticity of the DNA contained within our genes, and since there are many palindromic sequences near gene regulation sites, it has been postulated that these sequences may form hairpin structures during replication.[44,51,52] Thus, a DNA oligomer that forms a double hairpin might closely simulate natural occurrences.

Rentzeperis and co-workers[35] have probed the minor groove and the integrity of the base-pair stacking of the stem of unligated double hairpins, with a nick in the middle and lacking a phosphate group, by thermodynamically characterizing the binding of netropsin to a set of two double-hairpin molecules, with central helical sequences GGAT^TACC/GGTAATCC and CCAT^TAGG/CCTAATGG (both with four thymines in each loop) and their corresponding control core duplexes without the nicks, GGATTACC/GGTAATCC and CCATTAGG/CCTAATGG. Table 4 shows the resulting thermodynamic profiles. Netropsin binds to the central core of four A·T base pairs of all four sequences with similar binding affinities of about 10^8, consistent with the high specificity of netropsin for dA·dT base pairs. The similarity of the binding affinities indicates that the sequences of the binding sites that are contained in the duplexes, and in the unligated double hairpins with nicks, constitute identical binding sites. Thus, netropsin recognizes the A·T base pairs at the nick point as normal helical base pairs, that is, with similar base-stacking interactions and local helical parameters. One difference in the binding of netropsin to these two types of helical conformations is their response to salt concentration. The values of the slopes of ln K_b vs. ln [Na$^+$] are similar and equal to 1.1 for the duplexes and 1.0 for double hairpins; these low values are consistent with the lower charge density of all four duplexes. The small difference of 0.1 between the two types of duplexes may be explained in terms of the absence of a phosphate group at the nick point of the double hairpins. For these strong binding sites at the center of the molecules, we obtained similar thermodynamic profiles: ΔG_b^0 of -10.8 kcal mol^{-1}, $\Delta H_b^0 = -10.6$

Table 4. Thermodynamic Profiles for Netropsin Binding to Oligomer Duplexes[a]

Oligomer	Site	K_b (M^{-1})	ΔG_b^0 (kcal mol⁻¹)	ΔH_b (kcal mol⁻¹)	$T\Delta S_b$ (kcal mol⁻¹)	$d\ln K_b/$ $d\ln [Na^+]$
Duplex I	Stem	2.6×10^8 (2.6×10^7)	−10.7 (−9.4)	−10.6 (−10.8)	+0.1 (−1.4)	−1.1
Double Hairpin I	Stem	2.8×10^8 (4.2×10^7)	−10.7 (−9.7)	−10.6 (−10.4)	+0.1 (−0.7)	−1.0
	Loop	3.6×10^5 (5.8×10^4)	−7.1 (−6.1)	−4.7 (−3.4)	+2.4 (+2.7)	−0.9
Duplex II	Stem	3.6×10^8 (4.2×10^7)	−10.9 (−9.7)	−10.8 (−10.9)	+0.1 (−1.2)	−1.1
Double Hairpin II	Stem	2.9×10^8 (4.2×10^7)	−10.8 (−9.7)	−10.4 (−10.3)	+0.4 (−0.6)	−1.0
	Loop	1.1×10^5 (1.3×10^4)	−6.4 (−5.2)	−5.5 (−3.4)	+0.9 (+1.8)	−1.1

Notes: [a]Values were obtained from a combination of spectroscopic and ITC experiments in 10 mM sodium phosphate buffer, 0.1 mM Na2EDTA, pH 7.0, 5 °C (see Ref. 35 for details). The sequences used are d(CCATTAGG)/d(GGTAATCC) (duplex I); d(TAGGT4CCTAATGGT4CCAT) (double hairpin I); d(GGATTACC)/d(CCTAATGG) (duplex II); and d(TACCT4GGTAATCCT4GGAT) (double hairpin II).

kcal mol⁻¹, and $T\Delta S_b^0$ = +0.2 kcal mol⁻¹, with binding enthalpies independent of salt concentration. The ΔG_b^0 values decreased to −9.6 kcal mol⁻¹ with a 10-fold increase in NaCl concentration. In all cases these ΔG_b^0 terms correspond to binding processes that are primarily enthalpically driven. However, the decrease in the ΔG_b^0 is entropically driven when there is an increase in salt concentration and thus corresponds to a less favorable entropy term, $\Delta(T\Delta S_b)$ of −1.5 kcal mol⁻¹, consistent with processes driven by electrostatic effects.[38,53]

In the double hairpins, the presence of the thymine loops creates additional sites for netropsin, as detected in the calorimetric titrations, and each of these sites is able to accommodate two ligand molecules. In low-salt buffer, these sites are characterized with K_b's of about 10⁵ and exothermic ΔH_b's of −4.7 to −5.5 kcal mol⁻¹, and increase in salt concentration reduces the K_b's to about 10⁴ and ΔH_b's to −3.4 kcal mol⁻¹. The decrease in the

ΔG_b^0 value with increasing salt concentration is enthalpically driven ($\Delta\Delta H_b^0$ = 1.7 kcal mol^{-1}) and is due to the increase in hydrophobic interactions between two netropsin molecules in the same binding site. However, we obtained an average value of -1.0 mol of Na$^+$ per mol ligand from the slopes of ln K_b vs. ln [Na$^+$] plots for these sites, indicating that they have double-helical structure. The overall thermodynamic profiles for the association of netropsin to these secondary sites are similar to those of netropsin binding to GC sequences[8] and probably correspond to a nonspecific binding of netropsin. These secondary sites could contain the two dG·dC base pairs at each end of the double hairpins plus the adjacent two thymines of the loops, forming a local duplex structure of nearly three base-pair stacks. Alternatively, the constrained thymines of the loops could form these sites, as has been detected in the association of ethidium to the thymine loops of single hairpins.[29]

V. INTERACTION OF ETHIDIUM WITH NONCLASSICAL DNA STRUCTURES

A large array of different conformations of nucleic acid molecules has been uncovered and investigated throughout the years. Unusual DNA structures that are of special interest now are two sets of molecules: (1) telomeres that are contained in the ends of eukaryotic chromosomes, and (2) branched structures that may form transiently in the processes of DNA replication, recombination, and repair. The design of DNA model systems that takes into account the proper DNA sequence complementarity to mimic these multistranded structures has been very effective. For example, telomeres contain tandemly repeated clusters of guanine, and recently the in vitro properties of oligonucleotide models containing such clusters of guanine have been studied as models for these telomeric repeats,[54–58] as has the use of three- and four-arm immobile junctions to model branched structures.[59–63] The transient existence of these nonclassical structural states in DNA may raise fundamental questions: Are these states preferential targets for interaction of

Table 5. Thermodynamic Profiles for Ethidium Binding to
Multistranded Helices

Duplex	n	K_b (M^{-1})	ΔG_b^0 (kcal mol^{-1})	ΔH_b (kcal mol^{-1})	$T\Delta S_b$ (kcal mol^{-1})
G4-Tetramer[a]					
G4-DNA	1.0	1.5×10^5	−7.3	−5.9	+1.4
Control Duplex	2.7	5.2×10^4	−6.8	−12.0	−5.2
Three-Arm Junction[b]					
Y1	2.1	1.4×10^5	−6.9	−10.5	−3.6
	9.8	1.7×10^4	−5.7	−10.7	−5.0
Control Duplex	8.7	6.7×10^4	−6.5	−7.8	−1.3
Four-Arm Junction[b]					
J1	2.7	6.4×10^5	−7.8	−9.3	−1.5
	15.8	4.8×10^4	−6.3	−9.3	−3.0
Control Duplex	8.3	1.3×10^5	−6.9	−7.6	−0.7

Notes: [a]These values were obtained from the fit of calorimetric titrations in 10 mM sodium phosphate buffer, 0.1 mM Na$_2$EDTA, 200 mM NaCl, pH 7.0, 15.8 °C (see Ref. 64 for details).
[b]These values were obtained from the fit of calorimetric titrations in 20 mM sodium cacodylate buffer, 100 mM NaCl and 1 mM MgCl$_2$, pH 7.0, 20 °C (see Ref. 65).

ligands? How might these ligand interactions influence the recognition of such structures by proteins?

The thermodynamic profiles, resulting from the nonlinear fits of titration calorimetry binding isotherms, for the interaction of ethidium to three types of structures, G4 quartet[64] and three- and four-arm junctions,[65] are shown in Table 5. For proper comparison, the corresponding profiles of control duplexes with similar sequences and in identical solution conditions are also included. The experimental temperatures are slightly different and less than 16 °C for the G4-DNA set and 20 °C for the junction set.

A. Ethidium Interaction with Oligonucleotides Containing Tetraguanine Repeats

Qiu and co-workers analyzed the association of ethidium bromide with tetraplexes containing the sequence d(T$_4$G$_4$) and with the control duplex d(T$_4$G$_4$)/d(C$_4$A$_4$).[64] A tight binding interaction occurs, and the thermodynamic profile of the ethidium/tetraplex system (see Table 5), as well as the results of the observed spectral shifts and quantum yields, are consistent with an intercalative mechanism.[64] The $d(T_4G_4)$ tetraplex is able to accommodate only one ethidium molecule with a binding affinity of about 10^5, which is slightly higher than that of its control octameric duplex, which binds about three ligand molecules. In both cases, the association of ethidium is primarily enthalpy driven with a less favorable enthalpy contribution for the tetraplex structure. This heat contribution is lower than the overall heat of ethidium binding to dG·dC base pairs. The control duplex shows a much higher enthalpy of interaction that can be attributed to intercalation of ethidium between other types of base-pair stacks, such as TT/AA and/or TG/CT, that are present in this duplex. An alternative explanation is that the tight association of ethidium to the G-quartets, yielding a lower enthalpy contribution, is by a mechanism consistent with a partial stacking of the phenanthridine ring.

B. Ethidium Interaction with DNA Junctions

In the case of junctions with three and four helical arms containing a total of 24 and 32 base pairs, respectively, one would expect ethidium densities in these molecules of about 12 and 16 ligands per junction molecule, respectively, assuming that all the possible base-pair stacks are formed. For the control duplexes of 16 base pairs, we estimate an ethidium density of 7–8 ligands per duplex. The work of Hernandez and co-workers on this subject,[65] also included in the last two entry sets of Table 5, shows that the actual binding densities are 8–9 ligands per duplex for the control duplexes, 12 for the three-arm junction (Y1), and 19 for the four-arm junction (J1). However, there are two distinctive types of binding sites in each of these junction molecules. In Y1 this corresponds to 2 ligands characterized by a K_b of 1.4×10^5 M^{-1} and 10 ligands with a K_b of 1.7×10^4 M^{-1}, and in J1 this amounts to 3 ligands with a K_b of 1.3×10^5 and 16 ligands with K_b of 6.7×10^4 M^{-1}. The sites with stronger affinity correspond to the binding of ethidium to the crossover points in these structures and most likely are stabilized by the local and additional electrostatic contribution due to the higher charge density of phosphate groups at the center of these duplexes, as has been postulated earlier.[63] Ethidium binding to both types of sites takes place in enthalpy-driven processes with average exothermic heats of −9.3 kcal (Y1) and −10.6 kcal (J1). Comparison of these thermodynamic profiles with those of their respective control duplexes indicates an increased affinity for the strong sites, two-fold in Y1 and five-fold in J1, whereas the ligand affinity decreases for the weaker sites, by three-fourths in Y1 and by one-half in J1. Furthermore, the overall binding enthalpies are more favorable by −1.7 kcal (J1) and −2.8 kcal (Y1). The combined results yield the following observations: (1) the crossover points of both molecules constitute stronger binding sites for ethidium binding, and (2) on the average, ethidium binding is more favorable to four-arm junctions than three-arm junctions. One reason for this is that the base pairs that abut the crossover point of the three-arm junction are more perturbed, with these perturbations also extending to the helical arms causing more favorable heats of interaction. In the absence of counterion binding data, and because these experiments have the inclusion of magnesium ions, it is difficult to draw further conclusions from the binding entropies other than that these differences are larger.

VI. SUMMARY

Calorimetric techniques are powerful tools for the study of the energetics involved in the conformational plasticity of nucleic acids and in their interaction with ligands. The techniques of ITC and DSC, described here, provide us with the universal measurement of the heat associated in these reactions and allows dissection of the free energy terms into their corresponding enthalpic and entropic contributions. They also provide additional extrathermodynamic data such as, but not limited to, stoichiometry of complexes, ion binding, and cooperativity. In this review we have tried to give the reader a brief overview and discussion on how to apply calorimetric techniques for the understanding of ligand binding to DNA by using typical examples on the ligand mode of binding, ligand sequence specificity, the novel binding of ligands to hairpin loops, and the interaction of ligands to nonclassical DNA structures of current biological importance.

ACKNOWLEDGMENTS

This work was supported by Grant GM-42223 from the National Institutes of Health. The authors are grateful to Professor Louise Pape for critical reading and editorial changes of this manuscript.

REFERENCES

1. Johnson, P. F.; McKnight, S. L. *Ann. Rev. Biochem.* **1989**, *58*, 799–839.
2. Waring, M. J. *Ann. Rev. Biochem.* **1981**, *50*, 159–192.
3. Chou, W. Y.; Marky, L. A.; Zaunczkowski, D.; Breslauer, K. J. *J. Biomolec. Struct. Dynam.* **1987**, *5*, 345–359.
4. Marky, L. A.; Snyder, J. G.; Remeta, D. P.; Breslauer, K. J. *J. Biomolec. Struct. Dynam.* **1983**, *1*, 487–507.
5. Marky, L. A.; Blumenfeld, K. S.; Breslauer, K. J. *Nucleic Acids Res.* **1983**, *11*, 2857–2870.
6. Remeta, D. P.; Mudd, C. P.; Berger, R. L.; Breslauer, K. J. *Biochemistry* **1991**, *30*, 9799–9809.
7. Marky, L. A.; Snyder, J. G.; Breslauer, K. J. *Nucleic Acids Res.* **1983**, *11*, 5701–5715.
8. Marky, L. A.; Breslauer, K. J. *Proc. Natl. Acad. Sci. U.S.A.* **1987**, *82*, 4359–4663.
9. Bresloff, J. L.; Crothers, D. M. *J. Mol. Biol.* **1975**, *95*, 103–123.
10. Mandal, C.; Englander, S. W.; Kallenbach, N. R. *Biochemistry* **1980**, *19*, 5819–5825.
11. Wakelin, L. P. G.; Waring, M. J. *J. Mol. Biol.* **1980**, *144*, 183–214.
12. Ryan, D. P.; Crothers, D. M. *Biopolymers* **1984**, *23*, 537–562.
13. MacGregor, R. B., Jr.; Clegg, R. M.; Jovin, T. M. *Biochemistry* **1985**, *24*, 5503–5510.
14. Krugh, T. R.; Hook, J. W., III; Lin, S.; Chen, F. In *Stereodynamics of Molecular Systems*; Sarma, R. H., Ed.; Pergamon Press: NY, 1979; pp. 423–429.
15. Marky, L. A.; MacGregor, R. B., Jr. *Biochemistry* **1990**, *29*, 4805–4811.
16. Marky, L. A.; Kupke, K. J. *Biochemistry* **1989**, *28*, 9982–9988.
17. Patel, D. J.; Canuel, L. *Proc. Natl. Acad. Sci. U.S.A.* **1979**, *74*, 5207–5211.
18. Berman, H. M.; Neidle, S.; Zimmer, C.; Thrum, H. *Biochim. Biophys. Acta* **1979**, *561*, 124–131.
19. Patel, D. J. *Eur. J. Biochem.* **1979**, *99*, 369–378.

20. Patel, D. J. *Proc. Natl. Acad. Sci. U.S.A.* **1982**, *79*, 6424–6428.
21. Marky, L. A.; Curry, J.; Breslauer, K. J. In *Molecular Basis of Cancer, Part B: Macromolecular Recognition, Chemotherapy, and Immunology*; Rein, R., Ed.; Alan R. Liss, Inc.: New York, 1985; pp. 155–173.
22. Kopka, M. L.; Yoon, C.; Goodsell, D.; Pjura, P.; Dickerson, R. E. *Proc. Natl. Acad. Sci. U.S.A.* **1985**, *82*, 1376–1380.
23. Dervan, P. B. *Science (Washington, D.C.)* **1986**, *232*, 464–471.
24. Breslauer, K. J.; Ferrante, R.; Marky, L. A.; Dervan, P. B.; Youngquist, R. S. In *Structure & Expression, Volume 2: DNA & Its Drug Complexes*; Sarma, R. H., Ed., 1988; pp. 273–290.
25. Ward, B.; Rehfuss, R.; Goodisman, J.; Dabrowiak, J. C. *Biochemistry* **1988**, *27*, 1198–1205.
26. Coll, M.; Aymami, J.; van der Marel, G. A.; van Boom, J. H.; Rich, A.; Wang, A. H. *Biochemistry* **1989**, *28*, 310–320.
27. LePecq, J. B.; Paoletti, C. *J. Mol. Biol.* **1967**, *27*, 87–106.
28. Rentzeperis, D.; Kupke, D. W.; Marky, L. A. *Biopolymers* **1992**, *32*, 1065–1075.
29. Rentzeperis, D.; Alessi, K.; Marky, L. A. *Nucleic Acids Res.* **1994**, *21*, 2683–2689.
30. Rentzeperis, D.; Marky, L. A. *J. Am. Chem. Soc.* **1993**, *115*, 1645–1650.
31. Sturtevant, J. M. *Annu. Rev. Phys. Chem.* **1987**, *38*, 463–488.
32. Privalov, P. L.; Potekhin, S. A. *Methods in Enzymology* **1986**, *131*, 4–51.
33. Wiseman, T.; Williston, S.; Brandts, J. F.; Lin, L. N. *Anal. Biochem.* **1989**, *179*, 131–137.
34. Rentzeperis, D.; Karsten, R.; Jovin, T. M.; Marky, L. A. *J. Am. Chem. Soc.* **1992**, *114*, 5926–5928.
35. Rentzeperis, D.; Ho, J.; Marky, L. A. *Biochemistry* **1993**, *32*, 2564–2572.
36. Lin, L.-N.; Mason, A. B.; Woodworth, R. C.; Brandts, J. F. *Biochemistry* **1991**, *30*, 11660–11669.
37. Marky, L. A.; Breslauer, K. J. *Biopolymers* **1987**, *26*, 1601–1620.
38. Record, M. T., Jr.; Anderson, C. F.; Lohman, T. M. *Q. Rev. Biophys.* **1978**, *11*, 103–178.
39. Crothers, D. M. *Biopolymers* **1971**, *10*, 2147–2160.
40. Rentzeperis, D.; Kupke, D. W.; Marky, L. A. *J. Phys. Chem.* **1992**, *96*, 9612–9613.
41. Gillies, G. T.; Kupke, D. W. *Rev. Sci. Instrum.* **1988**, *59*, 307–313.
42. Chastain, M.; Tinoco, I., Jr. *Prog. Nucleic Acid Res. Mol. Biol.* **1991**, *41*, 131–177.
43. Draper, D. E. *Acc. Chem. Res.* **1992**, *25*, 201–207.
44. Maniatis, T.; Ptanshe, M.; Backman, K.; Kleid, D.; Flashman, S.; Jeffrey, A.; Maurer, R. *Cell* **1975**, *5*, 109–113.
45. Rosenberg, M.; Court, D. *Annu. Rev. Genet.* **1979**, *13*, 319–353.
46. Wells, R. D.; Goodman, T. C.; Hillen, W.; Horn, G. T.; Klein, R. D.; Larson, J. E.; Muller, U. R.; Neuendorf, S. K.; Panayotatos, N.; Stirdivant, S. M. *Prog. Nucleic Acid Res. Mol. Biol.* **1980**, *24*, 167–267.

47. Hahn, F. E. In *Antibiotics III;* Corcoran, J. W.; Hahn, F. E., Eds.; Springer-Verlag: New York, 1975, pp. 79–100.

48. Wartell, R. M.; Larson, J. E.; Wells, R. D. *J. Biol. Chem.* **1974**, *249*, 6719–6732.

49. Luck, G.; Triebel, H.; Waring, M.; Zimmer, Ch. *Nucleic Acids Res.* **1974**, *1*, 503–530.

50. Zimmer, C.; Wahnert, U. *Prog. Biophys. Molec. Biol.* **1986**, *47*, 31–112.

51. Muller, U. R.; Fitch, W. M. *Nature* **1982**, *298*, 582–585.

52. Murchie, A. I.; Alastair, I. H.; Bowater, R.; Aboul-Ela, F.; Fareed, A.; Lilley, D. M. J. *Biochim. Biophys. Acta* **1992**, *1131*, 1–15.

53. Manning, G. S. *Q. Rev. Biophys.* **1979**, *11*, 179–246.

54. Blackburn, E. H.; Szostak, J. W. *Annu. Rev. Biochem.* **1984**, *53*, 163–194.

55. Blackburn, E. H. *Nature* **1991**, *350*, 569–573.

56. Sen, D.; Gilbert, W. *Nature* **1988**, *334*, 364–366.

57. Williamson, J. R.; Raghuraman, M. K.; Cech, T. R. *Cell* **1989**, *59*, 871–880.

58. Sundquist, W. I.; Klug, A. *Nature* **1989**, *342*, 825–829.

59. Guo, Q.; Lu, M.; Churchill, M. E. A.; Tullius, T. D.; Kallenbach, N. R. *Biochemistry* **1990**, *29*, 10927–10934.

60. Lu, M.; Guo, Q.; Kallenbach, N. R. *Biochemistry* **1991**, *30*, 5815–5820.

61. Kallenbach, N. R.; Ma, R. I.; Seeman, N. C. *Nature* **1983**, *305*, 829–831.

62. Wang, Y.; Muller, J. E.; Kemper, B.; Seeman, N. C. *Biochemistry* **1991**, *30*, 5667–5674.

63. Seeman, N. C.; Kallenbach, N. R. *Annu. Rev. Biophys. Biomol. Struct.* **1994**, *23*, 53–86.

64. Guo, Q.; Lu, M.; Marky, L. A.; Kallenbach, N. R. *Biochemistry* **1992**, *31*, 2451–2455.

65. Hernandez, L. I.; Zhong, M.; Courtney, S. H.; Marky, L. A.; Kallenbach, N. R. *Biochemistry* **1994**, *33*, 13140–13146.

MOLECULAR MODELING OF DRUG–DNA INTERACTIONS:

FACTS AND FANTASIES

John O. Trent and Stephen Neidle

Advances in DNA Sequence Specific Agents
Volume 2, pages 29–58.
Copyright © 1996 by JAI Press Inc.
All rights of reproduction in any form reserved.
ISBN: 1-55938-166-3

I. INTRODUCTION

The study of DNA structure has always involved molecular modeling, with the original double-helix concept of Watson and Crick being the paramount example. It was developed by means of (by current standards) crude physical models and drawings in conjunction with the X-ray fiber diffraction data of Franklin and Wilkins. The elegance, power, and simplicity of their structure, and the fact that it has stood the test of time, are in large part due to the structure's not having been solely the result of a modeling exercise, but, critically, having been successfully tested against experimental diffraction data.[1]

Molecular modeling, even though it is now in many ways altogether more quantitative and stereochemically reliable, is still at its most powerful when it is closely allied to experiment, when it can sometimes rationalize and even successfully predict new findings. It is at its least useful, and thus open to most criticism, when it strays far from experimental approaches and attempts to exist in total isolation from them. Therefore, when applied at a level of theory appropriate to the problem in hand, modeling can be seen as a uniquely useful approach for the understanding and prediction of molecular behavior. The rise of molecular modeling from a highly specialized subject in the early 1980s to a now widely available set of computational and graphical tools is in large part due to the increasing availability of computer power, which, for a typical laboratory computer, has increased by a number of orders of magnitude in the past decade. Many of the methodologies in current use were developed prior to even the double-helix concept, but these have only relatively recently become realizable on molecular systems of more than trivial size.

The central problems common to all modeling studies on DNA are (1) its conformational complexity in terms of the number of torsional variables per repeating nucleotide unit and (2) the continuing difficulty of accurately dealing with the highly charged nature of nucleic acids, especially when they are interacting with other molecules. The mononucleoside building block has six backbone torsion angles to be considered together with the base–sugar glycosidic angle and flexibility in the deoxyribose sugar. Early models for DNA and drug–DNA structure tended to be based on the concept, taken from fiber diffraction analysis, that all the nucleoside units were conformationally identical. We now know, even within the restraints of a double helix, that this is a considerable oversimplification and that specific structural features are embedded within a sequence. Unfortunately, we are still a long way from being able to define the rules for sequence specific DNA structure, let alone understand how generalized sequences are recognized by large and small ligands.

Computerized molecular modeling studies on drug–DNA intercalation complexes were first reported in the early 1980s,[2–6] soon after several crystallographic analyses of dinucleoside–drug intercalation complexes.[7–9] These studies provided the first detailed experimental structural data of relevance to drugs interacting with biological DNA, albeit with obvious limitations imposed by the shortness of dinucleoside sequences. The modeling studies on these systems mostly used simple 6–12 van der Waals hard sphere intermolecular energy calculations, sometimes with no electrostatic contributions. Even so, they were able to reproduce successfully many of the geometric features observed experimentally,[10] especially drug–base stacking patterns. Calculations of intermolecular energetics were at times surprisingly successful in being able to qualitatively rationalize and rank observed drug–DNA binding affinities in terms of structural features. The subsequent development of much more sophisticated inter- and intramolecular force fields has enabled the problems of drug–DNA intercalation to be approached anew. However, therapeutic interest in intercalating agents has declined, in large part on account of their relative lack of selec-

tivity to tumor vs. normal cells, which is generally reflected at
the DNA level in their ability to bind to most sequences. Much
current activity is focused on sequence selective agents,[11] with
the hope that it will be possible to develop an understanding
of the rules governing sequence recognition, thereby making it
possible to eventually design molecules to recognize unique sequences in particular genes that have involvement in disease
states, especially human cancers. Over 100 discrete oncogenes
have now been identified, and these, together with a variety of
tumor-selective proteins, are potential targets for sequence specific agents.[12]

Computational modeling, in conjunction with computer graph-
ics visualization, has been used for (1) the rationalization and
prediction of conformation and stereochemistry, (2) the examination and prediction of sequence specificity, (3) the computation of thermodynamics and energetics of interactions, and (4)
the prediction of new analogues with enhanced DNA-binding
and/or sequence selectivity and (less frequently) the prediction
of entirely new DNA-binding compounds.

In this review, we shall survey the scope and limitations of
the principal modeling methods in current use and highlight
some of their applications to understanding drug–DNA interactions.

II. THE CURRENT TOOLS OF MOLECULAR
MODELING

A. Force Fields and Their Parameterization

The central problem that molecular modeling attempts to solve
is that of the calculation of the potential energy of a molecular
system with respect to the positions of all its component atoms.
In principle, this is described by the appropriate wavefunction
describing the system in quantum mechanical terms. However,
even very short oligonucleotides are far too large to be amenable
to meaningful quantum mechanical approaches. Instead, empiri-
cally derived force fields have been developed to calculate potential energies; such an approach has a long history in

theoretical chemistry, with empirical force fields having been used to study the stabilities of, for example, saturated hydrocarbons and simple peptides. The most frequently used force fields for both proteins and nucleic acids have the typical form[13,14]

$$E_{total} = \sum_{bonds} K_r(r - r_0)^2 + \sum_{angles} K\theta(\theta - \theta_0)^2 +$$

$$\sum_{torsion} (V_n/2)[1 + \cos(n\phi - \gamma)] +$$

$$\sum_{i<j} [A_{ij}/R_{ij}^{12} - B_{ij}/R_{ij}^6 + q_i q_j/\varepsilon R_{ij}] +$$

$$\sum_{\text{H-bonds}} (C_{ij}/R_{ij}^{12} - D_{ij}/R_{ij}^{10})$$

where the various terms describe contributions from bond length and bond angle deformations from equilibrium values, energy barriers to torsion angle changes, together with van der Waals, electrostatic, and hydrogen bond nonbonded interactions. This formalism is used by the two major force fields in common use for nucleic acids, CHARMM[13] and AMBER.[14] The hydrogen bond term is nondirectional, because it is widely considered that hydrogen bonds are largely electrostatic in origin and can be treated by appropriate choice of further van der Waals terms.

The intermolecular binding energy for a drug–DNA interaction can be defined from potential energy calculations of E_{total} on drug alone, DNA alone, and the drug–DNA complex as the energy of the complex minus the total energies of the drug and the DNA (in unperturbed, drug-free form)

$$\Delta E = E_{complex} - (E_{drug} + E_{DNA})$$

and DNA helix destabilization energy due to drug binding can be defined as

$$E_{destab} = (E_{DNA})_{complex} - (E_{DNA})_{native}$$

In order to describe completely a nucleic acid in force field terms, accurate values for the geometric parameters of all four common nucleosides and the phosphate unit are required, together with values for their force constants. The geometric parameters are in principle obtained from high-precision X-ray crystallographic analyses of nucleosides and nucleotides, together with microwave structures (where available). Information on force constants is generally obtained from the literature on infrared frequencies and normal mode analysis, though this is far from complete for nucleic acid constituents. Nonbonded parameters have been obtained from examination of lattice energies in simple model crystal structures. It is standard practice to assign atoms to particular types for assignment of these parameters, depending on their bonding environment. However, examination of a typical force field parameter set invariably shows that a significant number of force constants and nonbonded parameters are actually not available from experiment, but have to be assigned by analogy with established examples. Values for nucleic acid geometry can have small but significant variation from one parameter set to another. This is not surprising for the phosphodiester linkage, because this is not well represented by high-resolution X-ray structures. DNA bases, however, have been well studied and the variability here is less understandable. As yet there has been little systematic study of the effects of such variability on equilibrium geometries and energies.

Atom-centered charges on nucleic acid constituents have been calculated by use of ab initio quantum mechanically derived electrostatic potentials fitted to a point charge model. Atomic charges can also be obtained from very high resolution crystal structures (and hence reliable estimates of electron density), the valence occupancy being refined during the refinement process. This has been done for a number of mononucleosides and nucleotides,[15] and the results agree qualitatively with the calculated charges, albeit with a number of significant quantitative differences. For example, atom N-3 of adenine has an experimental charge of $-0.28e$, yet a calculated one of $-0.72e$. It may well be that the crystal environment has an effect on these experimental charges, and so their general use in force field calculations is probably premature. However, this does highlight the more general problem that calculated DNA charges are for isolated molecules, and the simple Coulombic term in the force field does not take account of the likely induced polarization effects of neighboring residues in a particular DNA conformation, yet alone of ligand- or solvent-induced polarization effects. Atom polarizability and nonadditive exchange repulsion terms have been included in an extension to the AMBER force field for calculations on the binding enthalpies of several ions with some crown ethers, with evident improvements in energies compared to experimental values.[16] There is a significant computational overhead in such calculations; presumably, this together with the lack of adequate parameterization of polarizability terms has so far hindered the use of these terms in biomolecular simulations.

In the absence of explicit solvent molecules in a modeling study, the treatment of the dielectric is of some significance: in a purely aqueous medium ε would have a value of 78.5, providing effective charge shielding at typical intermolecular distance. Thus, solvent-free molecular mechanics modeling should include a dielectric term that simulates this shielding effect. In practice, a simple distance-dependent dielectric can suffice, with $\varepsilon = 4r_{ij}$ producing stereochemically acceptable structural results.[17] A computationally more expensive sigmoidal function for the dielectric has been proposed but has not as yet found general favor.[18] Molecular dynamics simulations in the absence of explicit solvent use a value of $\varepsilon = 1$.

The problem of parameterization for DNA-binding ligands and drugs is a central one that needs to be satisfactorily resolved if reasonably realistic simulations and modeling on DNA complexes are performed, especially if reliable estimates of energies are required. On the other hand, sensible intermolecular geometries for DNA complexes can be obtained even with force field parameters (especially of force constants and torsion barriers) that are intelligent estimates. This is particularly the case when the starting model is close to the final one. So, for a given ligand, how can high-quality parameters be derived? Equilibrium geometries are obtainable from high-precision, small-molecule X-ray crystallographic analyses of the drug/ligand alone, with errors in individual bond lengths and angles being as small as ±0.002 Å and ±0.1° respectively. Averages of features from related structures taken from the Cambridge Crystallographic Database[19] of small-molecule structures will increase the reliability of individual values and may even be of use in suggesting plausible values for barriers to rotation for flexible regions of a molecule. Quantum mechanical calculations can provide electrostatic potential charges; high-quality semiempirical methods can, if carefully used, provide charge data of comparable quality to computational ab initio calculations, which are much more costly. They can also be used to calculate force constants, but analogy with existing values for comparable features should only be made with some care. An important pitfall to be avoided is the careless transfer of parameters from one force field to another without evaluation and testing, so that values are no longer self-consistent.

B. Molecular Mechanics and Dynamics Approaches

The total potential energy as defined by the force field equation above is an estimate of the internal energy of a system and is not the thermodynamic free energy measured experimentally. Thermodynamic energies can only be calculated in principle by methods that involve sampling a large number of configurational states. Molecular mechanics, which involves the minimization of the potential energy E_{total}, thus does not do so, even when explicit solvent molecules are included in the minimization. Molecular mechanics enables local potential energy minima to be sampled and optimized and has the advantages of simplicity and computational cheapness. It is most appropriate when a local minimum is well defined. Molecular dynamics,[20] by contrast, enables molecular motions to be simulated over a course of time by means of the application of Newtonian mechanics, usually using the same standard force field and parameterization as in molecular mechanics. Molecular dynamics then monitors the influence of kinetic energy and intra- and intermolecular forces on a set of atoms with respect to time and enables many local minima in a complex molecule such as a nucleic acid to be sampled. However, since each individual step of motion must be less than the period of highest frequency motion (typically one femtosecond), a total dynamics simulation for an oligonucleotide is typically only a few hundred picoseconds. The longest such simulation reported to date is of one nanosecond.[21] Thus, these simulations as yet do not approach chemical reaction or biological event timescales. Furthermore, molecular dynamics simulations of complex molecules, at least on these timescales, cannot fully explore all conformational space for large, flexible molecules and thus provide only pseudoglobal minimum information. The technique of simulated annealing, especially as implemented in the X-PLOR program,[22] is assuming considerable popularity both in purely modeling studies and in the refinement of incompletely resolved X-ray and NMR structures. It uses molecular dynamics methods in raising the temperature of a molecule to overcome local energy barriers and then gradually cooling it to arrive at "global" conformational minima.

It is possible, given sufficient supercomputer time, to systematically explore the full conformational space of simple DNA models such as dinucleosides. This approach has been pioneered by Broyde and Hingerty and their colleagues in studies of DNA–carcinogen covalent complexes.[23] In practice, it is necessary to limit searches to those domains already known to be preferred for nucleic acids. This restriction had the consequence that some significant global minima were not located in this study; by contrast, simulated annealing found more low-energy domains in fewer cpu hours.

C. The "Correctness" of Modeling: Crystal Structures as Test Beds?

How can the correctness of a molecular modeling study be judged? To a large extent this depends on the nature of the study itself and the questions it attempts to answer. The most apparently straightforward circumstance is that of modeling a structure, such as an AT-selective minor groove drug–DNA complex, in order to determine the optimal geometry of interaction. Given the DNA (i.e., an oligonucleotide) in a particular conformational class, such as B-form, even molecular mechanics calculations in the absence of solvent contributions should be capable of correctly locating the drug in an AT region of the DNA, with GC binding being of significantly higher energy. However, it is considerably more demanding for such calculations to distinguish finer detail, such as between different types of AT selectivity, for three principal reasons: (1) only local minima are sampled and so significant conformational change on binding cannot be located, (2) the absence of explicit solvent, and (3) deficiencies in the force field for both DNA and drug. Similar problems can arise with studies that attempt to define ranking orders of DNA binding for a series of ligands. However, when all such members of a series occupy roughly the same DNA-binding site, then the absence of solvent in the model may be a lesser problem, and it is possible to find good agreement between theory and experiment, at least in terms of qualitative ranking orders of binding. Since only potential energies are obtained from such calculations, the most that can be derived from them are intermolecular binding energies, which should not be formally compared with experimentally determined binding energies.

Can current explicit solvent molecular dynamics simulations and simulated annealing methods arrive at structures for drug–DNA complexes that compare well with experiment, in particular, with NMR or crystal structures? (The inclusion of explicit solvent molecules in simulations of nucleic acids and their drug complexes is generally accomplished by the use of a Monte Carlo procedure whereby water molecules are added to a structure, initially at random, and are then allowed to equilibrate to a physically acceptable arrangement. This procedure is computationally relatively expensive.) There are actually few instances where detailed comparisons have been made, but these do indicate that, given adequate simulation time to explore sufficient of the conformational plurality of a DNA structure, it is possible to reproduce at least some sequence-dependent features of the DNA, such as sugar pucker, propeller twist, and base-pair roll and possibly even groove width variations. Less success is achieved for localized solvent molecules that have particular involvement in the drug–DNA interaction itself. For example, the observation in several crystal structures of pentamidine derivative–oligonucleotide complexes[24,25] of a hydrophobic spine of hydration surrounding the drug has not been reproduced in simulations (Laughton, C. A., personal communication). It is likely that this is due to deficiencies in the simple TIP3P force field model currently used for water molecules.

Crystal structures of DNA–drug complexes are themselves products of molecular modeling methodologies. Crystallographic "refinement" is usually the least-squares optimization of a structural model with respect to the experimental observations of structure amplitudes. In general, the resolution of the diffraction data on a drug–oligonucleotide crystal structure is significantly less than atomic, with 2.0–2.5 Å being common. These crystal structures are underdetermined with respect to the number of independent atoms compared to the number of experimental observations; thus, refinements require constrained/restrained procedures in order to preserve acceptable geometries. The resulting structures are highly dependent on the care with which the refinement has been performed, as well as on the parameterization of such features as torsion barriers. Crystallographic refinements are increasingly performed with programs and parameter sets that are common to modeling and simulation studies generally (e.g., the use of the X-PLOR program[22] and its parameter set is especially widespread). Thus, any deficiencies in a parameter set can be inadvertently distributed across the otherwise distinct experimental techniques of crystallography and NMR. One should bear in mind that X-ray structures can be affected by crystal packing forces and should not be taken as the "perfect" bond distances and angles, just the best estimate.

D. Quantum Mechanical Approaches

Computer simulations[26] employing molecular mechanics and molecular dynamics together with Monte Carlo calculations are being increasingly used to study the structure and dynamics of DNA, to extend NMR studies in solution and X-ray single crystal studies. These simulations rely on the quality of the empirical parameters required to parameterize the force fields. The basic building blocks of DNA, the bases, must be studied at a very high level of theory in order to derive their parameters with an adequate level of reliability. Ideally, ab initio calculations on all molecules would be desirable. However, owing to the computational demands of quantum calculations, particularly ab initio, the upper limit for DNA calculations is currently on the order of a very few nucleotides.

The experimental trends of stacking and hydrogen bonding of DNA pairs has been examined at a high level of theory, including dispersion attraction. Gould and Kollman[27] have studied the solvent polarity effect on stacking versus hydrogen bonding of GC and AT Watson–Crick and Hoogsteen base pairs using SCF and MP2 calculations, and also utilizing a Dunning double ζ plus polarization basis set upon 6-31* SCF-optimized bases and base pairs. A number of calculations were carried out to determine which level of theory best reproduced the experimental energies. SCF HF/3-21G, HF/6-31G*, MP2/6-31G*, HF/3-21G, MP2/DZP//HF/6-31G*, HF/6-31G*//HF/6-31G*/, and HF/DZP//HF6-31G* with and without basis set superposition error correction (BSSE) were investigated. The bond lengths for the HF/3-21G and HF/6-31G* calculations agreed with experiment, whereas those from HF/MINI-1 were consistently too long. However, bond angles for the three techniques were in general agreement with experiment (only Hoogsteen AT base pairs can be directly compared with microwave experiment). The HF/3-21G basis set exaggerates the pairing energies, and consequently, it has been suggested that it should not be used for study of interaction energies. It was found, even at high levels of theory, that there was a single negative frequency for Watson–Crick and Hoogsteen AT base pairs, indicating a transition state, not a minimum, and further optimization was necessary. The HF/3-21G and HF/6-31G* levels gave good agreement with the experimental structures. The best estimates for energies of Watson–Crick GC and AT base pairs were at the MP2/DZP//6-31G*(BSSE) level of $-\Delta H_{298}$ of 25.4 and 11.9 kcal/mol, respectively. For the Hoogsteen AT base pair, the best estimate was $-\Delta H_{298} = 12.8$ kcal/mol (experimental mass spectrometry gives a value of 13 kcal/mol). However, the experimental value of 21 kcal/mol for a Watson–Crick GC base pair is somewhat lower than the calculated value. These calculations are the most extensive to date.

Free energy perturbation/molecular dynamics have been used to study[28] the solvation effect on the bases in vacuo and in water solution. The two most important categories of internal interactions in a DNA double helix are the hydrogen bonds and stacking between the bases. Various studies[27] have established that in nonpolar media hydrogen bonding is preferred, whereas in water, stacking is preferred. Many theoretical techniques have been applied to this area. Free energy perturbation allows the calculation of free energy changes for the processes $X_g + Y_g \rightarrow XY_g$, $Y_g \rightarrow Y_{aq}$, $X_g \rightarrow X_{aq}$, and $XY_g \rightarrow XY_{aq}$. These in turn allow the calculation of ΔG for the association processes in water. These calculations correctly predict the order and preferences of hydrogen bonding and stacking in nonpolar and aqueous solution. Stacking preferences in water are G/C > A/A > A/T (experimentally the G/C stacked pair is less stable than A/A and A/T). In nonpolar solvents the preference is G–C (hydrogen-bonded) > G/C (stacked) > A–T (hydrogen-bonded) > A/T (stacked) > A/A (stacked).

Earlier calculations[29] included the evaluation of the potential mean force to give a free energy of association as a function of the coordinate using coordinate coupling and the statistical perturbation theory of Zwanzig.[29] The gas phase and aqueous intermolecular potential for hydrogen-bonded and stacked bases, respectively, were calculated for adenine and thymine bases as a function of reaction coordinate. The minimum in the hydrogen-bonded potential is mainly due to electrostatics, whereas the stacked potential is mainly due to van der Waals interactions.

The stacked conformation was shown to be more favorable in solution and the hydrogen-bonded conformation more favorable in the gas phase. There is a shift in the potential mean force of 0.3 Å on going from gas phase to solution, which is accounted for by water destabilizing the hydrogen bonding of the base pairs.

Recently, the use of hybrid techniques has been investigated.[30] These techniques use a high level of theory, such as an ab initio or a semiempirical method, in the local area of interest, a reaction center or ligand binding site, and a lower theoretical level, such as a molecular mechanics force field, for the surrounding region. The principal difficulty in these hybrid techniques is how to deal with the interface between the two levels of theory. As increasing computational resources become available, this computationally intensive method will be more feasible, and this problem will undoubtedly be resolved.

III. NMR MODELING

The advent of multidimensional NMR techniques has made NMR an invaluable tool in the structural study of biological molecules. Although NMR is complementary to X-ray crystallography, with obviously different data collection and analysis strategies, the use of similar molecular modeling techniques is essential to both. A key difference between crystallography and NMR lies in the information obtained from them. Whereas DNA crystallography is often relatively close to atomic resolution and gives great detail on the overall structure, NMR gives very accurate local structural data only for parts of DNA. The drawback is that there is no long-range structural variable in NMR linking the detailed portions. Thus, molecular modeling is essential to determine the overall structure. NMR gives the jigsaw pieces; molecular modeling can give the whole picture, or more correctly, a series of pictures that are consistent with the pieces.

NMR information comes predominantly from proton nuclei, but additional information may be obtained with ^{13}C, ^{15}N, ^{17}O, and ^{31}P nuclei. There are three types of information obtainable for protons. The first is the NOE (nuclear Overhauser experiment), which yields interprotonic distances. The NOE gives through-space coupling, so essentially if protons are in proximity they will show an NOE peak with strength proportional to the interproton distance (strong is 2–2.5 Å, medium is 2.5–3.5 Å, and weak is 3.5–5 Å). The technique is not sensitive to interproton distances above 5 Å, as it relies on the spin–spin interaction of the protons involved. The second type of information obtainable for protons is on coupling constants (J). This is a through-bond interaction and gives information on bonding and, through the Karplus equation, dihedral angles of protons on adjacent carbons. The sugar rings in nucleic acids are usually well defined by the coupling constants. Third, it is possible to determine proton typing through chemical shift and to which heteroatom the proton is attached. The most useful nucleus is the proton, so most two-dimensional NMR on DNA is essentially proton mapping, but other nuclei can be informative. ^{31}P can be used for short sequences (6–14 base pairs) by full assignment, and the phosphate torsion angles can be obtained.[31] ^{13}C can be used in conjunction with ^{1}H to map the positions of protons on the sugars, and ^{15}N can be used to determine imine versus amine functionality.

In addition to static structural data, dynamic data can be derived from NMR. Phenomena that occur in the 10^3–10^9 sec^{-1} time range may be observed, such as ring flipping. Exchange rates with solvent (accessibility) may lead to the determination of some thermodynamic properties such as bound complex exchange at equilibrium and pH studies (including the determination of pK_a). Also, there is a greater probability that there are multiple conformations in solution than in the crystalline state, so the possibility exists of determining these, which may complicate the analysis.

A. Molecular Modeling Techniques Used in NMR

The fundamental assumption made in molecular modeling of DNA with NMR data is that the structure is basically known. Most techniques are followed by energy optimization using either the steepest descents or the conjugate gradient method.

The most common techniques and software (in brackets) used for NMR molecular modeling are listed below.[32]

1. Restrained molecular dynamics (rMD) (QUANTA, X-PLOR, BIOGRAPH, DISCOVER, GROMOS, AMBER).
2. Distance geometry (DGEOM, DGII, DISGEO, DSPACE, EMBED, VEMBED, XPLOR/dg).
3. Torsion space (DISMAN, DANA).
4. Grid searches.

Molecular dynamics is the most common molecular modeling technique applied in NMR and crystallography. The initial trial structure is allowed to evolve over time under the influence of a force field and Newton's equations of motion, with an additional equation to take into account the NOE data. Typically, in simulated annealing, the NOE constraints begin with low force constants and are gradually increased when the structure goes to elevated temperatures. The structure is then cooled and MD is continued with maximum NOE force constraints applied. The determination of a reasonable structure is usually governed by a penalty function regarding violations of the NOE data and torsion data.

$$E_{NOE} = k_{NOE}(d_{ij} - d_{ij}^*)_2$$

$$E_{tor} = k_{tor}(\Phi_{ij} - \Phi_{ij}^*)_2$$

The result of rMD is an ensemble of structures that satisfies the NOE data, being dependent on the quality of the force field and the parameters used in it. As already stated, the issue of parameterization of force fields used in molecular mechanics and dynamics in general is of central importance. To some extent the structure obtained depends on which force field is used, and so the structures derived from NMR are products of the NMR constraints and the force field. Distance geometry methods determine the structure using an ensemble of atom–atom distances and are usually in Cartesian space. Distance geometry starts with a matrix of distances that gives a distribution of structures, which are then further refined with penalty functions for distance violations and chirality errors. The determination of structure using torsional space, with bond lengths and angles being fixed, reduces the number of independent variables used in Cartesian space. Grid searches are only viable over small structures because they involve the systematic driving of the torsion angles. The computational overhead rapidly increases with the number of torsion angles and the driving resolution required.

B. An Example of Molecular Modeling with NMR: Microgonotropens

1, R = (CH$_2$)$_5$-N(CH$_2$CH$_2$CH$_2$-NMe$_2$)$_2$
2, R = (CH$_2$)$_4$-NH(CH$_2$CH$_2$)N(CH$_2$CH$_2$NH$_2$)$_2$

The minor groove noncovalently bound microgonotropens **1** and **2** in 1:1 complexes with d(CGCAAATTTGCG)$_2$ have been studied[33,34] with two-dimensional NMR and restrained molecular dynamics. The studies showed that **1** and **2** bind asymmetrically in the A–T region as 1:1 and 2:1 complexes and stiffen the DNA backbone. The native d(CGCAAATTTGCG)$_2$ structure was determined by two-dimensional NMR and restrained molecular dynamics and compared with the X-ray structure. It was also suggested, owing to observations of line broadening, that ligands **1** and **2** bind in a "flip-flop" mechanism of exchange to the two equivalent A$_3$T$_3$ binding sites.

The biological data are consistent with the modeled structures and agree on the effect of different substituents and the strength of binding of **1** and **2**. Molecule **2** is positioned deeper in the minor groove and exhibits a stronger interaction with the phosphate backbone than **1**, whereas neither **1** nor **2** deforms the DNA backbone greatly, as seen from the bending angles for the native solution structure above, **1** + d(CGCAAATTTGCG)$_2$, and **2** + d(CGCAAATTTGCG)$_2$ of 21.4°, 17.2°, and 22.2°, respectively.

1. Modeling on the Native d(CGCAAATTTGCG)$_2$ Structure

One hundred and two resonances were assigned and 78 NOEs were observed. From this, 66 distances were calculated and 47 distance constraints were used in rMD with the CHARMM[13] force field using the X-ray structure as the starting point. The minor groove in the solution structure narrows considerably between the T5–T9 and T4–T8 phosphates (from 3 to 1 Å, respectively) relative to the starting crystal structure, whereas the CHARMM minimization without constraints only narrowed the same regions by 0.3 to 1.0 Å. The helical bend in the solution structure (with constraints) is twice as large as in the crystal structure and in the minimized structure (no constraints), being 21.4°, 10.5°, and 7.5°, respectively. The other helical parameters of helical rise, axial rise, and turn angle were found to be similar in the crystal and solution.

2. Modeling on the Complex 1:d(CGCAAATTTGCG)$_2$

One hundred and ninety-six resonances were assigned and 105 intramolecular interactions were located (28 of which were previously used in the native structure); 21 intermolecular interactions were used in docking **1**. The pyrrole rings A and B are coplanar, whereas C is rotated 68° from coplanarity. Semiempirical calculations (AM1) indicate that this rotation is less favored by 2.5 kcal/mol (about the same as for a good H-bond). On comparison with the native dodecamer solution structure, the minor groove widens between the T5–T9 and T4–T8 phosphates (from 3 to 2 Å, respectively) on binding **1**. The bend in **1**:d(CGCAAATTTGCG)$_2$ (with constraints) is greater than in the X-ray structure, the minimized native structure (no constraints), or the minimized **1**:d(CGCAAATTTGCG)$_2$ complex (no constraints) but less than in the native solution structure, these angles being 17.2°, 10.5°, 7.5°, 9.8°, and 21.4°, respectively.

3. Modeling on the Complex 2:d(CGCAAATTTGCG)$_2$

One hundred and fifty-five intramolecular interactions were located, 17 were used to refine the previous native solution structure, and 17 were used in docking **2**. On comparison of the native and **2**:d(CGCAAATTTGCG)$_2$ solution structures, the minor groove widens between the T5–T9 and T4–T8 phosphates (from 3 to 4 Å, respectively). The bend in the complex structure is slightly greater than that in the native solution structure, being 22.2° and 21.4°, respectively.

IV. SOME CASE HISTORIES

A. Drug–DNA Intercalation

The intercalation process, whereby a planar aromatic chromophore becomes inserted between adjacent DNA base pairs, was initially characterized by a combination of biophysical methods (spectroscopic and hydrodynamic) and fiber diffraction/model building studies on polymeric DNA.[1] The latter enabled the intercalation concept to be translated into quasi-molecular terms, with emphasis being placed on the degree of unwinding induced at the intercalation site by the chromophore. Detection of unwinding was for some time considered to be the definitive method for classifying a molecule as an intercalator. More recently, it has become apparent that, although this effect is shown by drugs such as actinomycin, anthracyclines, many acridines, and ethidium, a significant number of molecules that bind to DNA and perturb its structure do not actually intercalate, even though they produce local unwinding. A more rigorous criterion for intercalation is that the base pairs at the site of interaction of the chromophore become separated by 6.8 Å rather than the 3.4 Å in B-form DNA.

Local extension of the DNA double helix so as to produce an intercalation site can be accomplished in a number of ways in terms of backbone conformational changes. Early model-building studies did not have any direct experimental conformational data to use as a starting point, so it is not surprising that these show a wide variety of backbone conformations for intercalated DNA.[2–6,35] This is perhaps suggestive of there being a large number of energetically equivalent ways to form an intercalation site. On the other hand, crystallographic studies on dinucleoside–drug complexes[7–9,35] showed a single category of backbone conformation. The relevance of these experimentally derived models to intercalation into longer sequences is questionable, since dinucleosides have the obvious limitation of not providing information on the conformational effects of intercalation beyond the immediate base pairs of the binding site. Nonetheless, the dinucleoside–drug complexes have provided a useful starting point for modeling studies on structurally simple intercalating drugs that have focused on, for example, chromophore–base interactions. More recent extensive X-ray crystallographic studies on anthracycline–hexanucleotide complexes[36] have provided structural data of significantly greater relevance to the intercalation of these drugs into long DNA sequences. However, even in these structures (as well as those of sequence specific bisintercalators) end effects need to be taken into account, because the drug molecules are invariably intercalated close to the ends of hexamer duplexes. Despite this inherent limitation, molecular modeling studies, even without full geometry optimization, can successfully reproduce the known sequence preferences for these drugs.[37] Drug intercalation into an embedded site in a sequence of sufficient length (ten base pairs) so as to minimize end effects has been simulated by explicit solvent molecular dynamics.[38] This has shown that the conformational effects of forming an intercalation site can be propagated in several base pairs on either side of the site itself.

The crystal structure of actinomycin complexed to the self-complementary sequence d(GAAGCTTC) is the first to provide direct structural data on an intercalating drug embedded within a DNA duplex and not bound at or close to the ends.[39] This complex crystallizes in three different space groups,[40] with a number of conformational features appearing to vary between the three structures, although the moderate resolution of 3.0 Å limits their reliability. In all cases, the drug is intercalated at the G-3',5'-C site in a sequence specific manner, in accord with the extensive data on this drug. Features of the observed backbone conformation, such as a *trans,trans,trans* geometry for one strand of the phosphodiester 3' to the intercalation site, are unexpected. These experimental studies have been used in studies aimed at modeling the intercalation process itself.[40] An ingenious extension of the crystal structures has been explored via AMBER-based molecular modeling[41] in order to predict possible modifications to the actinomycin molecule that would facilitate binding to DNA:RNA hybrids such as are involved in the replication phase of retroviruses. The N-8 derivative of actinomycin was predicted to form a strong hydrogen bond to a 2'-hydroxyl of an RNA strand. Solution binding data suggest that although this compound does, in contrast to the parent drug, bind to a DNA:RNA hybrid as well as to a DNA:DNA one, overall its strength of binding is much reduced. This is thus an example of a case where, even though the modeling is based on a crystal

actinomycin

structure, molecular mechanics calculations cannot reliably predict even the ranking order of binding strengths. Here the absence of solvent contributions may be a factor, but more
probably the details of drug derivative–DNA conformation are
sufficiently distinct from the starting crystal structure so as to
render the modeling of only limited use.

The need to model with experimental structural data being
taken into account is illustrated by recent studies on a complex
between a decanucleotide and the enediyne antitumor antibiotic
esperamicin. Solvated molecular dynamics simulations over a
300-ps timescale, normally considered to be a long timescale,
resulted in models that placed the methoxyacrylyl–anthranilate
functionality in the DNA major groove.[42] On the other hand,
two-dimensional NMR in conjunction with restrained molecular
dynamics calculations with distance restraints unequivocally
places this moiety in the minor groove.[43]

B. Sequence Specific Drug Binding

Minor-groove-binding drugs interact with DNA with minimal
change in the DNA backbone, even with covalent minor groove
binders. The drugs[44] interact through alkylation or bind noncovalently with shape and charge complementarity with the walls
of the minor groove.

1. Covalent Minor-Groove-Bound Drugs

Naturally occurring pyrrolo[1,4]benzodiazepines (PBDs), such
as tomaymycin[45] and anthramycin,[46] reversibly alkylate DNA
at the N-2 position of guanine residues with a preference for
purine neighboring bases. Molecular modeling has been used[45–
47] to determine stereochemical (R/S at the site of alkylation)
and directional (pointing to 3' or 5') preferences. The stereochemical preference has been found[48] to vary with the PBD and
sequence, although seldom with an R predominance. The sequence selectivity of anthramycin is for d(purine-G-purine)
sites.

Site of alkylation

general pyrrolo[1,4]benzodiazepine

tomaymycin

anthramycin

A predictive modeling study[49] has been reported on the design of DSB-120, a C-8-linked pyrrolo[1,4]benzodiazepine dimer, and this study has also predicted its sequence selectivity. The alkyl A–A ring linker joins two PBD units that intermolecularly cross-link d(GXXC)$_2$ sequences. The modeling predicted a high preference for spanning bases d(GATC)$_2$, which has been experimentally confirmed.[50,51]

DSB-120

2. Noncovalent Minor-Groove-Bound Drugs

Repeat units of aromatic or pyrrolo amides have an AT selectivity, as shown by molecules such as distamycin,[52] netropsin,[53] Hoechst 33258,[54] and the previously mentioned
microgonotropens. These noncovalent minor groove drugs make
use of electrostatic interactions, hydrogen bonding, and van der
Waals interactions to stabilize the binding.

Complexes of distamycin and DNA have been widely studied.
A detailed molecular dynamics investigation[55] found that distamycin stiffens B-form DNA and stops any interconversion to
the A-form. The electrostatic contribution to the binding energy
of distamycin to d(CGCGAAATTTCGCG)$_2$ was found to be
slightly greater than the van der Waals contribution. Distamycin
can also bind as a side-by-side dimer in the minor groove of
AT-rich sequences that include at least some GC base pairs (with
a wider minor groove). Molecular modeling from NMR data
has shown that the base selectivity may be extended by use of

1-methylimidazole-2-carboxamidenetropsin

distamycin

2-imidazoledistamycin

derivatives of distamycin or netropsin, such as 1-methylimida-zole-2-carboxamidenetropsin[56] and 2-imidazoledistamycin,[57] to form side-by-side antiparallel dimers.

There are few examples[58] of free energy perturbation theory being applied to DNA–drug complexes, as it is computationally challenging, owing to the size of the system and the difficulties in adequate treatment of the electrostatic interactions. The binding energies of distamycin and 2-imidazoledistamycin have been studied[59] by use of molecular dynamics and free energy perturbation theory. This analysis correctly predicted the order of stability established by experiment and suggests that the method is reliable enough to play a major role in future quantitative simulation studies.

3. Hybrid Molecules

The aim of developing sequence selective agents is to be able to target any DNA sequence. One approach is to have an array of subunits that bind specifically to short sequences, which then can be easily linked together. Unfortunately, we are some way from this situation at present, although the side-by-side dimer approach has considerable promise. Molecular modeling will undoubtedly play a major role in the design of these subunits. Attempts have been made to combine CG and AT selectivity with a monocationic lexitropsin[60] using the JUMNA methodology, although the sequence selectivity was not high. This methodology has also been used[61] for a netropsin–oxazolopyridocarbazole com-

lexitropsin

netropsin–oxazolopyridocarbazole

bined minor groove binder and intercalating agent; this shows a high poly[d(AT)] preference.

V. SCOPE AND CURRENT LIMITATIONS

We have endeavored to show in this review that molecular modeling, covering as it does a wide range of approaches, has much to offer in the drug–DNA field. However, a note of caution is needed. It is tempting for the novice (and sometimes even the experienced worker) to simply use the latest plug-and-go graphics/computational black-box package. Such an approach can only produce meaningful results if one appreciates what can and cannot be done at the present time:

- the level of theory to be used in a particular problem should always be appropriate to the quality and type of data available, and to the type of answers required

- current molecular mechanics force fields generally handle geometries well and predictably and are especially useful for simple drug design
- parameterizations for DNA are now at a high level of reliability
- the derivation of force field parameters for individual drugs requires considerable effort
- calculation of thermodynamic quantities by free energy perturbation methods is now feasible, although they are computationally extremely expensive
- we still have a relatively poor understanding of sequence-dependent DNA structural features, in large part because of the lack of experimental data on a sufficient number of sequences. This is reflected in the inability of empirical force fields to adequately handle these features.

ACKNOWLEDGMENTS

We are grateful to the Cancer Research Campaign for support, various colleagues at the Institute of Cancer Research for discussions, and to Suse Broyde for preprints in advance of publication.

REFERENCES

1. Neidle, S. *DNA Structure and Recognition;* Oxford University Press: Oxford, UK, 1994.
2. Nakata, Y.; Hopfinger, A. J. *Biochem. Biophys. Res. Commun.* **1980**, *95*, 583–588.
3. Ornstein, R. L.; Rein, R. *Biopolymers* **1979**, *18*, 1277–1291.
4. Pullman, B. *Adv. Drug Res.* **1989**, *18*, 1–113.
5. Lybrand, T.; Kollman, P. *Biopolymers* **1985**, *24*, 1863–1879.
6. Islam, S. A.; Neidle, S.; Gandecha, M.; Partridge, M.; Patterson, H.; Brown, J. R. *J. Med. Chem.* **1985**, *28*, 857–864.
7. Neidle, S.; Achari, A.; Taylor, G. L.; Berman, H. M.; Carrell, H. L.; Glusker, J. P.; Stallings, W. C. *Nature* **1977**, *269*, 304–307.
8. Shieh, H.-S.; Berman, H. M.; Dabrow, M.; Neidle, S. *Nucleic Acids Res.* **1980**, *8*, 85–97.
9. Sobell, H. M.; Tsai, C.-C.; Jain, S. C.; Gilbert, S. G. *J. Mol. Biol.* **1977**, *114*, 333–365.
10. Islam, S. A.; Neidle, S. *Acta Crystallogr.* **1984**, *B40*, 424–429.

11. Lown, J. W. In *Molecular Aspects of Anticancer Drug–DNA Interactions;* Neidle, S., Waring, M. J., Eds.; Macmillan Press: London, 1993; Vol. 1, pp. 322–355.

12. Neidle, S.; Puvvada, M. S.; Thurston, D. E. *Eur. J. Cancer* **1994**, *30A,* 567–568.

13. Brookes, B. R.; Bruccoleri, R. E.; Olafson, B. D.; States, D. J.; Swaminathan, S.; Karplus, M. *J. Comp. Chem.* **1983**, *4,* 187–217.

14. Weiner, S. J.; Kollman, P. A.; Case, D. A.; Singh, U. C.; Ghio, C.; Alagona, G.; Profeta, S.; Weiner, P. *J. Am. Chem. Soc.* **1984**, *106,* 765. Weiner, S. J.; Kollman, P. A. *J. Comp. Chem.* **1986**, *7,* 230–252.

15. Pearlman, D. A.; Kim, S.-H. *J. Mol. Biol.* **1990**, *211,* 171–187.

16. Howard, A. E.; Singh, U. C.; Billeter, M.; Kollman, P. A. *J. Am. Chem. Soc.* **1988**, *110,* 6984–6991.

17. Orozco, M.; Laughton, C. A.; Herzyk, P.; Neidle, S. *J. Biomolec. Struc. Dynam.* **1990**, *8,* 359–373.

18. Lavery, R.; Sklenar, H.; Zakrzewska, K.; Pullman, B. *J. Biomolec. Struc. Dynam.* **1986**, *3,* 989–1014.

19. Allen, F. H.; Kennard, O.; Watson, D. G.; Brammer, L.; Orpen, A. G.; Taylor, R. *J. Chem. Soc., Perkin Trans. II* **1987**, S1–S19.

20. Van Gunsteren, W. F.; Berendsen, H. J. C. *Angew. Chem.* **1990**, *29,* 992–1023; Beveridge, D. L.; Ravishanker, G. *Current Biology* **1994**, *4,* 246–255.

21. McConnell, K. J.; Nirmala, R.; Young, M. A.; Ravishanker, G.; Beveridge, D. L. *J. Am. Chem. Soc.* **1994**, *116,* 4461–4462.

22. Brünger, A. T. In *Molecular Dynamics;* Goodfellow, J. G., Ed.; Macmillan Press: London, 1991; pp. 137–178.

23. Shapiro, R.; Sidawi, D.; Miao, Y.-S.; Hingerty, B. E.; Schmidt, K. E.; Moskowitz, J.; Broyde, S. *Chem. Res. Toxicol.* **1994**, *7,* 239–253.

24. Nunn, C. M.; Jenkins, T. C.; Neidle, S. *Biochemistry* **1993**, *32,* 13838–13843.

25. Nunn, C. M.; Jenkins, T. C.; Neidle, S. *Eur. J. Biochem.* **1994**, *226,* 953–961.

26. Pranata, J.; Jorgensen, W. L.; *Tetrahedron* **1991**, *47(14),* 2491–2501. Stewart, E. L.; Foley, C. K.; Allinger, N. L.; Bowen, J. P. *J. Am. Chem. Soc.* **1994**, *116,* 7282–7286.

27. Gould, I. R.; Kollman, P. A. *J. Am. Chem. Soc.* **1994**, *116,* 2493–2499.

28. Cieplak, P.; Kollman, P. A. *J. Am. Chem. Soc.* **1988**, *110,* 3734–3739; Kollman, P. A. *Chem. Rev.* **1993**, *93,* 2395–2418.

29. Dang, L. X.; Kollman, P. A. *J. Am. Chem. Soc.* **1990**, *112,* 503–507.

30. Field, M. J.; Bash, P. A.; Karplus, M. *J. Comp. Chem.* **1990**, *1,* 700–733.

31. Gorenstein, D. G. *Chem. Rev.* **1994**, *94,* 1315–1338.

32. *NMR of Macromolecules;* Roberts, G. C. K., Ed.; Oxford University Press: Oxford, 1993. Kuntz, I. D.; Thompson, J. F.; Oshiro, C. M. *Methods in Enzymology;* Oppenheimer, N. J., James, T. L., Eds.; Academic Press: San Diego, 1989; Vol. 177B, pp. 159–204. Wüthrich, K. *Methods in Enzymology;* Oppenheimer, N. J., James, T. L., Eds., 1989; Vol. 177B, pp. 125–131, Academic Press, San Diego.

33. Blasko, A.; Bruice, T. C. *Proc. Natl. Acad. Sci. U.S.A.* **1993**, *90,* 10081–10122.

34. Blasko, A.; Browne, K. A.; He, G.-X.; Bruice, T. C. *J. Am. Chem. Soc.* **1993**, *115*, 7080–7092.
35. Berman, H. M.; Neidle, S.; Stodola, R. K. *Proc. Natl. Acad. Sci. U.S.A.* **1978**, *75*, 828–832.
36. Wang, A. H.-J. In *Molecular Aspects of Anticancer Drug–DNA Interactions;* Neidle, S., Waring, M. J., Eds.; Macmillan Press: London, 1993; Vol. 1, pp. 32–53.
37. Chen, K. X.; Gresh, N.; Pullman, B. *Mol. Pharmacol.* **1986**, *30*, 279–286.
38. Herzyk, P.; Goodfellow, J. M.; Neidle, S. *J. Biomolec. Struc. Dynam.* **1992**, *10*, 97–139.
39. Kamitori, S.; Takusagawa, F. *J. Mol. Biol.* **1992**, *225*, 445–456.
40. Kamitori, S.; Takusagawa, F. *J. Am. Chem. Soc.* **1994**, *116*, 4154–4165.
41. Chu, W.; Kamitori, S.; Shinomiya, M.; Carlson, R. G.; Takusagawa, F. *J. Am. Chem. Soc.* **1994**, *116*, 2243–2253.
42. Langley, D. R.; Golik, J.; Krishnan, B.; Doyle, T. W.; Beveridge, D. L. *J. Am. Chem. Soc.* **1994**, *116*, 15–29.
43. Ikemoto, N.; Kumar, R. A.; Dedon, P. C.; Danishefsky, S. J.; Patel, D. J. *J. Am. Chem. Soc.* **1994**, *116*, 9387–9388.
44. Gago, F.; Reynolds, C. A.; Richards, W. G. *Molec. Pharm.* **1989**, *35*, 232–241.
45. Boyd, F. L.; Stewart, D.; Remers, W. A.; Barkley, M. D.; Hurley, L. H. *Biochemistry* **1990**, *29*, 2387–2403.
46. Rao, S. N.; Singh, U. C.; Kollman, P. A. *J. Med. Chem.* **1986**, *29*, 2484–2492.
47. Remers, W. A.; Mabilia, M.; Hopfinger, A. J. *J. Med. Chem.* **1986**, *29*, 2492–2503.
48. Thurston, D. E. In *Molecular Aspects of Anticancer Drug–DNA Interactions;* Neidle, S., Waring, M. J., Eds.; Macmillan Press: London, 1993; Vol. 1, pp. 54–88.
49. Jenkins, T. C.; Hurley, L. H.; Neidle, S.; Thurston, D. E. *J. Med. Chem.* **1994**, *37*, 4529–4537.
50. Smellie, M.; Kelland, L. R.; Thurston, D. E.; Souhami, R. L.; Hartley, J. A. *Br. J. Cancer* **1994**, *70*, 48–53.
51. Mountzouris, J. A.; Wang, J.-J.; Thurston, D.; Hurley, L. H. *J. Med. Chem.* **1994**, *37*, 3132–3140.
52. Coll, M.; Frederick, C. A.; Wang, A. H.-J.; Rich, A. *Proc. Natl. Acad. Sci. U.S.A.* **1987**, *84*, 8385–8389.
53. Marky, L. A.; Breslauer, K. L. *Proc. Natl. Acad. Sci. U.S.A.* **1987**, *84*, 4359–4363.
54. Kumar, S.; Joseph, T.; Singh, M. P.; Bathini, Y.; Lown, J. W. *J. Biomolec. Struct. Dynam.* **1992**, *9*, 853–880. Fede, A.; Billeter, M.; Leupin, W.; Wüthrich, K. *Structure* **1993**, *1*, 177–186.
55. Boehncke, K.; Nonella, M.; Schulten, K.; Wang, A. H.-J. *Biochemistry* **1991**, *30*, 5465–5475.
56. Geierstanger, B. H.; Jacobson, J. P.; Mrksich, M.; Dervan, P. B.; Wemmer, D. E. *Biochemistry* **1994**, *33*, 3055–3062.

57. Geierstanger, B. H.; Dywer, T. J.; Bathini, Y.; Lown, J. W.; Wemmer, D. E. *J. Am. Chem. Soc.* **1993**, *115*, 4474–4482.
58. Härd, T.; Nilsson, L. *Nucleosides and Nucleotides* **1991**, *10*, 701–709.
59. Singh, S. B.; Ajay; Wemmer, D. E.; Kollman, P. A. *Proc. Natl. Acad. Sci. U.S.A.* **1994**, *91*, 7673–7677.
60. Randrianarivelo, M.; Zarkrzewska, K.; Pullman, B. *J. Biomolec. Struct. Dynam.* **1989**, *6*, 769–779.
61. Goulaouic, H.; Carteau, S.; Subra, F.; Mouscadet, J. F.; Auclair, C.; Sun, J. *Biochemistry* **1994**, *33*, 1412–1418.

X-RAY CRYSTALLOGRAPHIC AND NMR STRUCTURAL STUDIES OF ANTHRACYCLINE ANTICANCER DRUGS:

IMPLICATION IN DRUG DESIGN

Andrew H.–J. Wang

Advances in DNA Sequence Specific Agents
Volume 2, pages 59–100.
Copyright © 1996 by JAI Press Inc.
All rights of reproduction in any form reserved.
ISBN: 1-55938-166-3

I. INTRODUCTION

Many antitumor/anticancer compounds exert their cytotoxic activities by interfering with the function of nucleic acids and of
proteins that interact with nucleic acids. Extensive research on
these compounds, which may be of synthetic or natural origin,
has revealed that there are five major types of DNA-binding
antitumor drugs, namely, intercalator, noncovalent groove
binder, covalent binding/cross-linking agent, DNA cleaving
agent, and nucleoside analogue.[1-3] Each type of these potent
agents targets specific DNA sequences. Understanding the molecular aspects of the drug binding is of particular interest, as
they may be relevant for the design of better drugs. The three-
dimensional structure of the complexes between antitumor drugs
and DNA oligonucleotides, determined by either single crystal
X-ray crystallography or NMR spectroscopy, offers the first step
in the design process. In this chapter, we first provide an overall
view of these two important techniques and then review examples of the three-dimensional structures of several anthracycline
drug–DNA complexes determined by these methods.

We chose anthracycline drugs as our representative systems
because the anthracycline antibiotics, such as daunorubicin
(DNR) and doxorubicin (DOX) (Figure 1), constitute an important family of widely used clinical anticancer drugs.[1-3] The principal cellular targets for these important drugs are presumed to

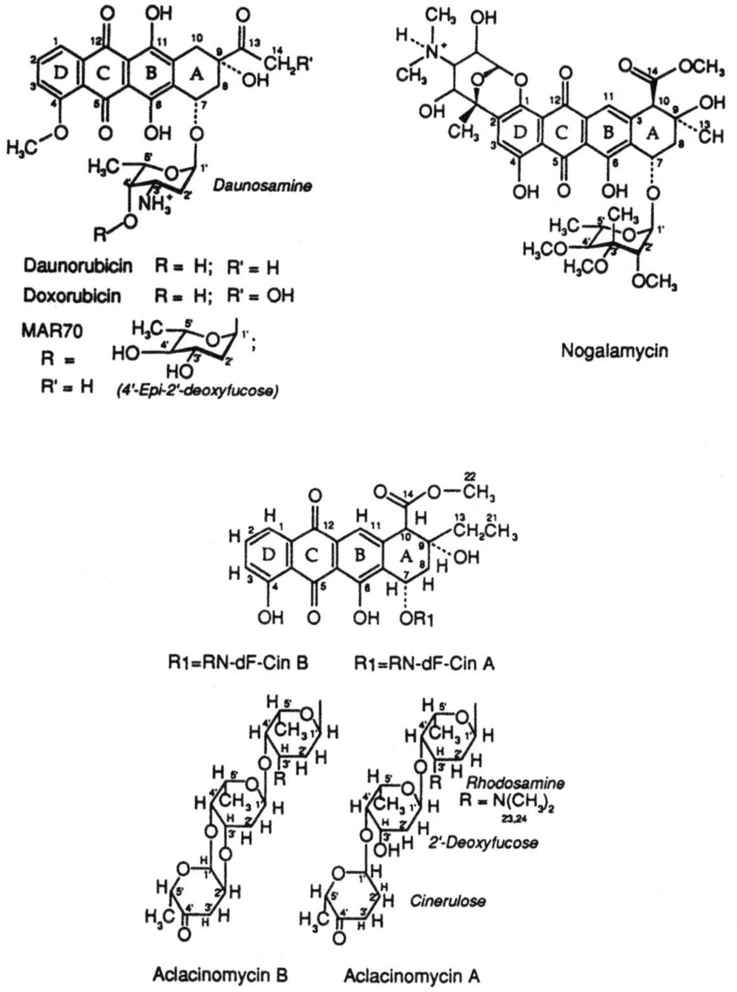

Figure 1. Molecular structures of anthracycline drugs described in this chapter. All molecules contain an aglycone chromophore with four fused rings (A–D). Different sugars are attached to the aglycone at the C-7 position.

be the DNA in chromosomes. Despite the widespread use of these drugs, their effectiveness is often hampered by their significant cardiotoxicity and the drug resistance that develops in some cancer cells. Recent progress in rational drug design offers a new opportunity in dealing with these problems. Rational drug design requires a full understanding of how anthracycline drug molecules interact with their DNA receptor (e.g., the binding affinity and specificity toward DNA), that is, a comprehensive study of the structure–function relationships of the existing drugs. Toward this goal, structural analyses of several anthracycline drug–DNA complexes have provided valuable information regarding the role of various functional groups of the drug molecules.[4]

It has been shown that although drug intercalation into DNA via the aglycone chromophore is a requirement for the DNA binding of anthracycline antibiotics, and most probably is needed for their biological activities, it is not a sufficient requirement, since the aglycone alone is not an active anticancer agent. Components other than the intercalator chromophore play essential roles in determining whether the compounds possess antitumor activity. Presumably, these components contribute by endowing on the compounds different DNA-binding affinities, DNA sequence specificities, or other required properties, such as membrane transport characteristics. In addition, they may affect the ways in which proteins (e.g., polymerases, helicase, and topoisomerases) interact with the drug–DNA complexes.[1–3] Indeed, many new anthracycline derivatives, involving modifications in either the aglycone or the sugar moieties, have been synthesized and tested for biological activities. The structural analyses of the drug–DNA complexes allow us to gain insights into the roles of various functional components in these compounds.

II. X-RAY CRYSTALLOGRAPHY

X-ray diffraction involves the interaction of X-rays with electrons in molecules. In a crystal, molecules are arranged in a highly regular and repetitive manner to form the lattice, and the electrons of molecules in the crystal lattice scatter X-rays collectively. A diffracted beam (reflection) can be detected when the incident X-ray beam (with a wavelength λ) glances the Miller planes [interplanar spacing $d(hkl)$] at an angle, θ, resulting in a reflected beam exiting at the same angle due to the constructive interference of the X-rays in the path. The Bragg equation depicts this condition, i.e., $n\lambda 2d(hkl)\sin\theta$.

The goal of X-ray diffraction analysis is to obtain a correct electron density map from which an initial molecular model can be constructed and then refined. The electron density distribution of a molecule in the crystal lattice at a position (x,y,z), $\rho(x,y,z)$, is expressed by the following equation:

$$\rho(x,y,z) = \frac{1}{V} \sum_h \sum_k \sum_l |F(hkl)| \, e^{-i[2\pi(hx+ky+lz)-\Phi(hkl)]}$$

Here $|F(hkl)|$ and $\Phi(hkl)$ are the magnitude and the phase angle, respectively, of the structure factor of a reflection. These two quantities need to be determined experimentally in order to calculate the electron density map. The process of structural analysis by X-ray diffraction involves several steps, which are concisely discussed below and more extensively described elsewhere.[5]

A. Crystallization

Crystallization of nucleic acids and their complexes with anticancer drugs remains somewhat empirical. Nevertheless, significant progress on crystallization methods has been made. Pure DNA oligonucleotides can now be prepared routinely on any automated DNA synthesizer to yield sufficient amounts of material for crystallization. Specific chemical modifications have been useful during either crystallization or subsequent structure determination. For example, molecules selectively substituted with a thiophospho linkage in the sugar–phosphate backbone seem to facilitate crystallization under certain conditions. Likewise, the iso-structural 5-bromodeoxyuridine can be used to replace thymidine in DNA as a heavy atom derivative to obtain multiple isomorphous replacement phase information.

The most commonly used crystallization technique is the vapor diffusion method described by McPherson.[6] Several factors, such as pH, metal ions (e.g., sodium, magnesium), and spermine, play important roles in the crystallization experiment. In addition, heavy metal ions such as Zn^{2+}, Co^{2+}, Ni^{2+}, and cobalt(III) hexaammine are also useful in promoting crystallization. A popular precipitating agent in the crystallization is 2-methyl-2,4-pentanediol (2-MPD). Another organic precipitant that may be used is a low molecular weight polyethylene glycol (PEG), such as PEG 400. Volatile organic solvents, such as isopropanol, have also occasionally been used. Several special techniques, such as feeding and seeding methods, are useful in increasing the size of the crystals. Commercial crystallization kits are now available from sources such as Hampton Research Inc. (Riverside, CA).

B. Data Collection

Crystals are mounted in a capillary along with a droplet of the mother liquor to prevent the loss of solvent molecules from the crystal lattice, for data collection. The capillary is placed on a goniometer head, and the crystal is centered in the X-ray beam on either a diffractometer or an area detector system. Typical X-ray sources are from rotating anode X-ray generators, which produce X-rays from a copper K_α (Cu K_α) line with a wavelength of about 1.54 Å. More recently, the use of synchrotron radiation has become more common.

The diffracted intensity, $I(hkl)$, is related to the structure factor, $F(hkl)$, with $I(hkl) = k \cdot Lp \cdot |F(hkl)|^2$. Here k is a correction factor for absorption, crystal decay, etc., and Lp is a geometric/physical correction factor (Lorentz polarization factor). It is clear that the structure factor amplitude, $|F(hkl)|$, can be obtained, but the phase information, $\Phi(hkl)$, of the reflection is lost during the data collection and can only be recovered via indirect means.

C. Solving the Phase Problem

The most critical step in X-ray crystallography is to recover the lost phase information of the diffraction data, that is, the solution of the phase problem. For molecules with fewer than 100 non-hydrogen atoms, the so-called direct method is invaluable; however, the size of the drug–DNA complexes is beyond the power of the direct method at present.

The classical method is the multiple isomorphous replacement technique, in which a series of heavy-atom-substituted isomorphous crystals are prepared. Heavy atom derivatives may be prepared either by substitution of the thymine base with a 5-bromouracil base or by soaking the native crystals with heavy metal ions such as Co^{2+} and Ba^{2+}. These modified crystals have very similar unit cell dimensions and the same space group, but their diffraction patterns are slightly perturbed (compared to that of the native crystal) because of the presence of heavy atoms in the lattice. The diffraction data of both the native (N) and heavy atom (H) derivative crystals are collected, providing their corresponding structure factor amplitudes $|F_N|$, $|F_{NH1}|$, $|F_{NH2}|$, etc.

The structure factor of a heavy atom derivative NH1 may be expressed vectorially as

$$\mathbf{F}_{NH1} = \mathbf{F}_N + \mathbf{F}_{H1}$$

or

$$|F_N| \exp(i\Phi_N) = |F_{NH1}| \exp(i\Phi_{NH1}) - F_{H1}$$

F_{H1} is calculated from the positions of the heavy atom H1, which can be located by the method of the difference Patterson map (a special type of Fourier map using $|F(hkl)|^2$, instead of $|F(hkl)|$, as the coefficient in Fourier synthesis). This allows two possible solutions for the phase angle of F_N. Additional heavy atom derivatives, NH2, NH3, etc., in combination with the first derivative (NH1) will afford an unique solution of F_N (with some uncertainties due to the experimental errors) so that an electron density Fourier map may be calculated.

The other widely used technique is the molecular replacement method. A model of the molecule, which is assumed to be very similar to the real structure, is rotated in the unit cell to determine the possible orientation of the molecule. The model in the preferred orientation is then systematically translated along the three directions (x, y, z) of the unit cell in order to find the correct position of the molecule in the crystal. Hence, the method is often called the rotation–translation search method. Several useful computer program packages, including X-PLOR,[7] have been developed for this method.

Finally, a new method, termed multiple anomalous dispersion, has been successfully applied to several protein structures. This method relies on the tunability of the synchrotron radiation source, which allows the anomalous scattering data to be collected at several different wavelengths. This method has not been widely used in the analysis of drug–DNA complexes.

D. Model Building

The resulting electron density map, $\rho(x,y,z)$, calculated using $|F_o|$ and Φ_N, is plotted to visualize the molecule. Several computer programs, including FRODO and O can display the electron density maps as a density envelope. The map plus the model can be manipulated interactively to optimize the fit of the model to the density. Figure 2 shows an example of one section of such mini maps of the DNR–CG(araC)GCG complex.[8]

E. Refinement

The initial model structure is used to calculate the theoretical structure factor, $|F_c|$, which is not in full agreement with the observed experimental structure factor, $|F_o|$. The initial model requires many adjustments, or refinements, so that $|F_c|$ and $|F_o|$ are in better agreement. Crystals of small molecules (with fewer than 100 atoms) often diffract X-rays to high resolution, e.g., 0.8 Å. This resolution allows the accuracy of small-molecule structure to be very high, with an estimated standard deviation (esd) of bond lengths of 0.002 Å. The very small esd's

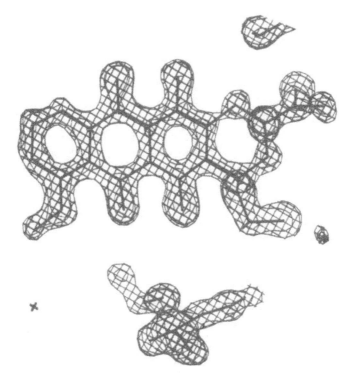

Figure 2. Electron density map from the crystal structure of a drug–DNA complex. The aglycone ring of the daunorubicin in the DNR–CG(araC)CGCG complex is clearly visible.

are the consequence of an important characteristic of the X-ray diffraction analysis, namely, the overdeterminacy.

The concept of overdeterminacy may be illustrated using the Z-DNA crystal structure of d(CGCGCG), which crystallizes in the orthorhombic space group $P2_12_12_1$ with the unit cell dimensions of $a = 17.9$ Å, $b = 30.98$ Å, and $c = 44.57$ Å.[9] There are two DNA hexamer strands, two spermines, 1 magnesium ion, and 68 water molecules, totaling 340 non-hydrogen atoms, in the asymmetric unit of the unit cell. Diffraction data up to 0.9-Å resolution ($2\theta = 125°$ using Cu K_α X-rays) were collected with 14,000 (out of 21,000 possible) observed reflections above $2\sigma(F_o)$ of the background. In order to refine the structure, three positional parameters (x,y,z) fixing the positions and six thermal parameters (temperature factors) describing the anisotropic motions of every individual atom in the molecule are needed. There are a total of 3060 (340 × 9) independent parameters to be determined. Therefore, in this crystal structure, there are 4.6 observations per variable parameter to be determined, that is, a 4.6-fold overdeterminacy.

However, many crystals of drug–DNA complexes tend to diffract X-rays to lower resolution (e.g., 3 Å); consequently, there are fewer available observed reflections. The reduced reflections: parameter ratio (underdeterminacy) requires a different refinement procedure in which individual atomic least-square refinement cannot be used. This involves a stereochemically constrained refinement, which substantially reduces the number of parameters to be refined by a factor of about 5. It becomes possible to refine a protein or DNA structure even at relatively low resolution (e.g., ~3 Å). In this process, instead of the individual atomic positions being refined, torsional angles are varied by least-square minimization of the function Ω_{total} ($\Omega_{total} = \Omega_F + \Omega_{distance} + \Omega_{vdW} + \Omega_{planar} + \Omega_{chiral} + \Omega_B$). Here the various terms to be minimized are the structure factors (Ω_F), bonding distances ($\Omega_{distance}$), nonbonded van der Waals contacts (Ω_{vdW}), planar groups such as the peptide bond or nucleic acid bases (Ω_{planar}), stereoconfiguration of chiral centers (Ω_{chiral}), and temperature factors (Ω_B). More recently, a new structure factor restrained refinement procedure incorporating molecular dynamics has been shown to be very effective. The simulated annealing procedure, as described in X-PLOR,[7] appears to be powerful in getting the model to the global minimum energy state.

The initial model of macromolecular crystals inevitably does not account for everything in the unit cell. The remaining material in the unit cell (e.g., solvent molecules and ions) can be located through the difference Fourier electron maps by use of the following equation:

$$\Delta\rho(x,y,z) = \frac{1}{V} \sum_h \sum_k \sum_l (|\,|F_o(hkl)| - |F_c(hkl)|\,|)$$

$$\times e^{-i[2\pi(hx + ky + lz) - \Phi_c(hkl)]}$$

Here $|F_o|$ and $|F_c|$ are the observed experimental and the calculated structure factors, respectively, and Φ_c is the calculated phase angle. This method is particularly useful in locating the missing portion of the model.

The progress of the refinement may be monitored by use of the crystallographic residual factor (R factor, $R = \Sigma\,|\,|F_o| - |F_c|\,|/\Sigma|F_o|$), which measures the agreement (or discrepancy) between the $|F_o|$ and $|F_c|$. For small molecules, one can now routinely attain an R factor of below 5%, because of the very reliable measurement of the intensity data. In contrast, the crystals of macromolecules contain large solvent channels that make it difficult to account for the presence of disordered solvent molecules. In addition, some regions of these large molecules are flexible, which contributes to the uncertainty of the model. Usually a macromolecular structure refined to an R factor of 15–20% with good constraints can be considered satisfactory.

It is important to point out that the R factor alone should not be the only criterion for judgment of the correctness of the structure. One should inspect the electron density map very carefully to see whether there is any obvious discrepancy between the model and the electron density in the map. Otherwise, a misinterpretation of the structure may occur.

III. NMR SPECTROSCOPY

Nuclear magnetic resonance (NMR) spectroscopy is another powerful technique used in elucidation of the three-dimensional structure of biological macromolecules. Its application for determination of nucleic acid structure has been discussed in previous volumes and shall not be repeated here, except for some recent progress, which will be briefly mentioned. In particular, very high field NMR spectrometers (> 750 MHz) are now available that provide a very high sensitivity such that biological samples (proteins or nucleic acids) can be studied at low concentration (0.1 mM or lower). Various multidimensional NMR techniques have been developed, although the traditional two-dimensional nuclear Overhauser enhancement spectroscopy

(2D-NOESY) and *J*-correlated spectroscopy (2D-COSY) are still the mainstays in structural determination. They provide information regarding the through-space and through-bond nuclear spin connectivity, respectively.

In a two-dimensional NMR experiment, an additional time domain (evolution domain, usually called the t_1 domain) is incorporated between an initial pulse (normally a $\pi/2$ pulse) and the second pulse (another $\pi/2$ pulse) before the acquisition of the free induction decay in the second time domain (t_2 domain). These two time domains are converted by Fourier transformation into their respective frequency domains, F_1 and F_2, which are related to the chemical shifts of the protons. In the NOESY experiment there is an additional $\pi/2$ pulse and delay (τ_m), termed the mixing time. During the mixing-time period, the protons in the molecule have an opportunity to cross-relax among themselves, resulting in the exchange of spins. This cross-relaxation is known as the nuclear Overhauser enhancement (NOE) and is observed as the off-diagonal cross-peaks in the 2D-NOESY spectrum.

The rate (ρ_{ij}) of the cross-relaxation between protons i and j is inversely proportional to sixth power of the distance between the two protons, r_{ij}, that is,

$$\rho_{ij} \propto r_{ij}^{-6} \cdot F(\tau_c)$$

where $F(\tau_c)$ is a function of the rotational correlation time of the molecule. This inverse sixth power of the distance dependence of these cross-relaxation rates makes the measured NOE very weak when the distance between two protons is > 5 Å. The observed cross-peak intensity results from multiple pathways for relaxation within the mixing-time period. At a very short mixing time, the observed NOE cross-peak volume is dominated by the largest term, that is, the direct relaxation between the two spins. However, as the mixing time becomes longer (> 25 ms), many secondary relaxation pathways become significant (or even dominant) and a direct interpretation of the cross-peak volume becomes unreliable. Thus, at longer mixing times, the cross-peak intensities are interpreted within the context of a relaxation matrix, encompassing the rate of relaxation between all nuclei simultaneously.

A. Analysis

In a typical NMR analysis, every proton resonance in the spectrum must first be assigned. A common method is sequential assignment, which allows one to establish the connectivity pathway along the linear backbone of the molecule. Different sugar and glycosyl conformations in the nucleic acid molecule place their protons in a unique arrangement, which are revealed in their respective COSY and NOESY spectra. In B-DNA the H-8 proton of a purine or the H-6 proton of a pyrimidine is not only close to its own H-2' and H-2'' protons, but also close to the H-1' and H-2'' protons of the 5'-side preceding deoxyribose. Indeed, strong NOEs are observed between those protons, and they may be used to follow the connectivity along the sugar–phosphate backbone of DNA. In the end, the full resonance assignment of the NOESY spectrum provides the entry point for obtainment of the entire dataset of the volumes of all NOE cross-peaks.

A critical step in elucidation of the solution structure of biological macromolecules is to reliably measure the NOE volumes of all the cross-peaks associated with any given proton. However, many cross-peaks of the NOESY spectrum are severely overlapped. In order to calculate the volume of each cross-peak, the overlapped peaks have to be deconvoluted. We have developed an approach, called SPEDREF (SPEctral-Driven REFinement), in which the deconvolution function of an overlapped region is estimated on the basis of the calculated NOE volumes derived from a molecular model.[10] The theoretical NOEs are calculated using the full-matrix relaxation theory.[11] The advantage of this approach is that in principle only the NOE spectrum at a single mixing time (e.g., τ_m = 200 msec) needs to be measured, substantially reducing the amount of time for collecting the NOE data. In practice, more than one NOE spectrum is collected at different mixing times.

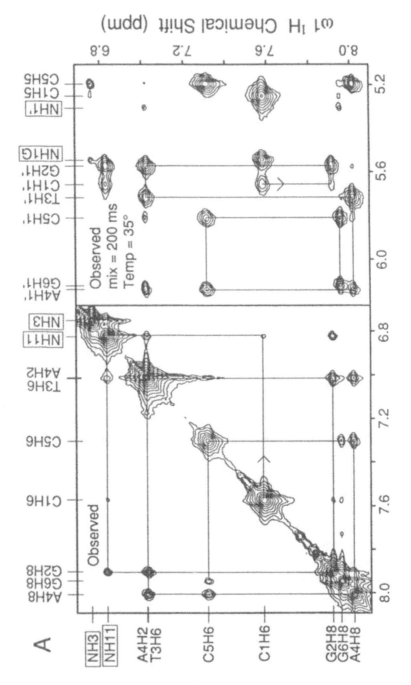

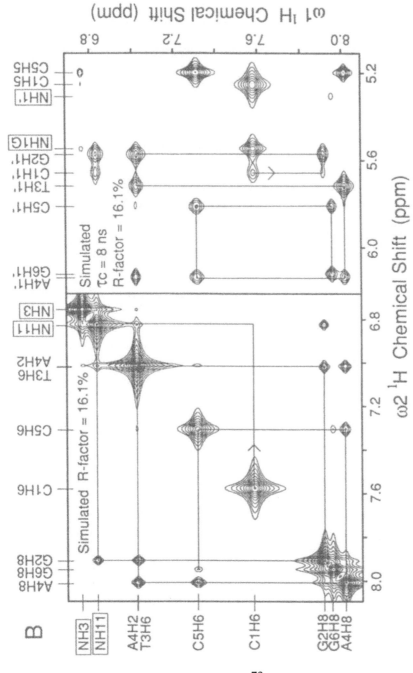

Figure 3. Experimental and simulated 2D-NOESY spectra of the 2:1 nogalamycin–CGTACG complex. The aromatic protons to all nonexchangeable protons cross-peak region is shown.

73

In the NMR refinement, these measured rates, ρ_{ij}, are used, along with the molecular correlation time (τ_c), to define inter-nuclei distances for constraints to be used by a refinement program, such as the energy minimization/molecular dynamics program X-PLOR.[7] The resulting minimized model from the X-PLOR refinement becomes the new starting model for the calculation of the deconvolution function for the next cycle of refinement. The process is iterated many times until convergence. The progress of the refinement is monitored by an NMR discrepancy factor [R factor(NMR) $= \Sigma \, | \, N_o - N_c \, | \, / \Sigma \, | \, N_o \, |$], analogous to the crystallographic R factor. Here the N_o and N_c are the respective observed and calculated off-diagonal NOE volumes. Figure 3 shows an example of a comparison between the experimental and simulated 2D-NOESY spectra of the 2:1 nogalamycin–CGTACG complex.[12]

Additional information, such as the sugar pucker conformation derived from the J coupling constants, may be added as constraints in the refinement procedure. The vicinal coupling constants may be used to calculate the torsion angles of the C-1′–C-2′ bond by use of the modified Karplus equation. Since the H–C–C–H torsion angles (e.g., H-1′–C-1′–C-2′–H-2′) are closely related to the sugar pucker conformation, they have been used to evaluate the conformational state of the sugar moiety (e.g., C-3′-*endo*-ribose in RNA or C-2′-*endo*-deoxyribose in DNA) of the nucleic acid backbone. Similarly, the coupling constants between the proton (e.g., H-5′/H-5″) and ^{31}P may be measured by use of the heteronuclear COSY method. These coupling constants may be used to define the backbone conformation of the phosphodiester linkage. Recent development in three- and four-dimensional NMR spectroscopy permits additional dimensions (e.g., ^{13}C and ^{15}N) to be added to resolve highly overlapped regions, and this is especially useful for studies on RNA molecules.

B. Dynamic Information

An important application of NMR spectroscopy to nucleic acids and their complexes with antitumor drugs is the study of the dynamic aspects of their binding interactions. In solution, the molecules under study are in constant motion (translation, rotation, vibration, etc.). The timescale for the vibrational motions is too short to be detected by NMR technique; instead, they may be studied by infrared or Raman spectroscopies. The molecular picture presented by the NMR technique is an averaged view of a large ensemble of conformations in rapid equilibrium. Some of these motions affect the line shape of the spectrum. For example, if the molecule is sufficiently large, its tumbling motion (which dominates the correlation time) becomes slow, which would result in a broad spectrum. With an increase in the temperature of the solution, molecules will tumble faster, resulting in sharper resonances. The slow chemical exchange between different molecular species is frequently observed in the systems of drug–DNA complexes. This is easily detected when the ratio of the drug to its DNA binding sites is less than stoichiometric. We will discuss this in Sections VI and VII.

NMR has also been used to determine the exchange rate of protons in nucleic acids. The exchangeable amino and imino proton resonances are located in the 7–9- and 12–14-ppm chemical shift ranges, respectively. The exchange of the imino protons with the solvent water molecules requires that the base pair open up transiently owing to the structural fluctuation of the double helix. The rate of the imino proton exchange in B-DNA has been determined to be in the range of milliseconds. This method is applicable to drug–DNA interactions, which provide useful information regarding the binding kinetics of the interaction.

IV. DAUNORUBICIN AND DOXORUBICIN AND THEIR DERIVATIVES

The three-dimensional crystal structure of the 2:1 complex between DNR and CGTACG at atomic resolution provided the first detailed view of how the anthracycline drug molecules bind to DNA.[13] In the complex, the aglycone chromophore intercalates between the CpG steps at both ends of a distorted B-DNA

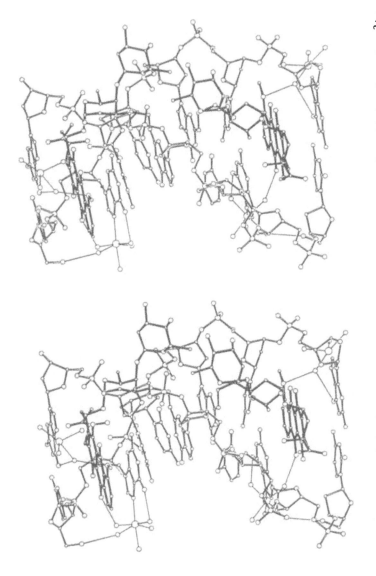

Figure 4. Stereoscopic skeletal drawing of the 2:1 MAR70–CGCGCG complex with four bound Co^{2+} ions.

double helix. The elongated aglycone (rings A–D) penetrates the DNA double helix with its D ring protruding into the major groove and the amino sugar lying in the minor groove.

The above and other subsequent structural analyses allowed us to identify three major functional components of anthracyclines: the aglycone intercalator (rings B–D), the anchoring function associated with ring A, and the sugars. Figure 4 shows the detailed interactions as exemplified by the structure of the complex of MAR70 and CGCGCG in the presence of a Co^{2+} ion.[14–16] The structure clearly shows that each component of the drug molecule plays a different, important role in the biological activity of the drug. The aglycone, by intercalating into DNA, causes a distortion in the double helix that may be recognized by enzymes (e.g., polymerases, topoisomerase II, topoisomerase I, helicase). The sugars located in the grooves of DNA are essential for additional interactions with relevant enzymes. The O-9 hydroxyl in the DNR–DOX series provides key hydrogen bonds to DNA, anchoring the drug firmly in the double helix and favoring a guanine base sequence. Finally, the configuration at the C-7 position in ring A is important, because it joins the amino sugar to the aglycone with a right-handed chirality so that the drug can position the amino sugar in the minor groove of a right-handed B-DNA double helix. In the following we summarize some structural studies on the DNR–DNA and DOX–DNA complexes.

A. DNA Binding Sequence

Several anthracycline–DNA complexes with different DNA sequences have been studied to address the question of the sequence specificity of DNR and DOX. The high-resolution structures of the DNR–CGATCG and DOX–CGATCG complexes[17,18] showed that most of the structural features seen in the DNR–CGTACG complex are preserved.[13] However, there is an interesting sequence dependence on the binding of the amino sugar to the AT base pair outside the intercalation site. In the DNR/DOX–CGATCG complexes, there are additional direct hydrogen bonds between the positively charged N-3′ amino group in the sugar and the O-2 of both C_{11} and T_{10} residues of DNA.[17,18] This suggests that DNR and DOX may bind more readily to the 5'-CGA sequence than to the 5'-CGT sequence. This DNA base triplet specificity agrees with the prediction from solution[19] and theoretical studies.[20] Further studies using the DNR–TGATCA and DNR–TGTACA complexes provide additional information on the sequence-dependent effect of DNR binding.[21] It was suggested that 5'-TGA is a better site than 5'-TGT, a similar sequence-dependent preference as listed above. Several recent crystal structures essentially augment the earlier information.[22,23]

An interesting observation about the structures is that the N-3' amino group of the sugar in the drug molecule approaches the edge of the base pairs in the minor groove. In the DNR–$C_1G_2T_3A_4C_5G_6$ structure, N-3' is 3.29 Å from O-2 of the C_5 cytosine base, 3.39 Å from O-4' of C_5 ribose, 3.44 Å from N-3 of A_4 adenine, and 3.52 Å from C-2 of A_4.[13] The former three distances may be considered as very weak hydrogen bonds. If there is a guanine at the fourth (and the symmetry-related tenth) sequence position in a hexamer such as $C_1G_2C_3G_4C_5G_6$, the two amino groups (N-3' from DNR and N-2 of G_4) in the nonadduct complex would be in close contact, which might slightly destabilize the binding of DNR to a sequence of 5'-GCG in DNA. Again, this is consistent with the sequence specificity of the binding of DNR to DNA mentioned above, i.e., 5'-(A/T)CG over 5'-GCG. In this case, these two amino groups (N-2 from guanine and N-3' from DNR) are brought into close proximity and are made rigid in the somewhat hydrophobic environment of the minor groove by the intercalative binding of DNR to DNA. This affords an ideal situation for an enhanced nucleophilic attack on amino groups by an agent such as formaldehyde (HCHO). In fact, the crosslinking reaction by HCHO transforms the unfavorable contact between the two amino groups existing in the noncovalently bonded complex to the covalent insertion of a methylene group. A more detailed discussion of this is presented in Section V.

B. Modifications in the Aglycone

An important issue in the clinical use of the anthracycline drugs is their cardiotoxicity.[1-3] The aglycone chromophore has been implicated in the biochemical processes that generate the free radical form of the aglycone, leading to the toxic effect. By modification of the aglycone (e.g., 4-*O*-demethyl and 11-deoxy), the modified anthracyclines may become better drugs.

Recently, X-ray diffraction analyses of the complexes between three aglycone-modified anthracyclines, 11-deoxy-DNR with d[CGT(pS)ACG][24] and idarubicin (4-demethoxy-DNR, IDR) and 4-*O*-demethyl-11-deoxy-DOX (ddDOX) with CGATCG,[25] provided the detailed three-dimensional molecular structures. Although the overall structures of these complexes are similar to those of DNR–DNA and DOX–DNA complexes, the missing C-4 methoxy of IDR and the missing methyl group at the O-4 position of ddDOX result in a different binding surface in the major groove for possible protein recognition. The O-4 hydroxyl group is capable of receiving and/or donating a hydrogen bond to proteins that bind to the drug–DNA complex. The missing O-11 hydroxyl group in ring B of the ddDOX–CGATCG complex creates an empty space in the intercalation cavity between the two GC base pairs, which appears to affect the stacking interactions between the aglycone and the DNA. These structural changes in the major groove of the drug–DNA complexes due to the modifications of the aglycone may be responsible in part for the difference in their biological activities.

C. Modifications in Daunosamine

Some derivatives of DNR and DOX with modifications in the aminosugar have more desirable biological properties. Epirubicin (4'-epi-DOX) and esorubicin (4'-deoxy-DNR) are third-generation synthetic anticancer agents that have been tested for clinical use. In epirubicin the orientation of the O-4' hydroxyl group is reversed compared to that of DNR. In the structure of the DNR/DOX–DNA complexes, the O-4' hydroxyl group is found to project out toward the solvent region. The crystal struc-

ture of epirubicin bound to CGATCG was solved,[26] and it
showed that the O-4' hydroxyl is hydrogen-bonded to the N-3
of the adenine base. This additional hydrogen bond makes the
epirubicin bind more effectively than doxorubicin. Another derivative, esorubicin, has been crystallized with several DNA hexamers, and their structures have been analyzed (unpublished
results). In these structures, the O-4' hydroxyl is missing, which
makes the sugar in the minor groove slightly more hydrophobic.
A different modification is the addition of another ring system
to either the O-4' or N-3' position. MAR70 is a derivative with
a sugar attached to the O-4' position. The structural analysis
of MAR70 complexed to CGTACG, d(CGTDCG), and
CGCGCG revealed that the second sugar projects into the solvent region, away from the bottom of the minor groove.[14–16]
The structure of the MAR70–CGCGCG complex is shown in
Figure 4.

V. COVALENT ADDUCTS OF DNR AND DOX WITH DNA

We recently discovered that HCHO can cross-link DNR to certain sequences of DNA very efficiently.[15,16] We have shown by
HPLC and X-ray diffraction analyses that when DNR is mixed
with CGCGCG in the presence of HCHO stable covalent adducts
of DNA are formed. These adducts contain a covalent methylene
bridge between the N-3' of daunosamine and the N-2 of a guanine. The reason for this efficient cross-linking reaction is the
perfect juxtaposition of two amino groups in the minor groove,
as described previously. The inserted methylene bridge does not
perturb the conformation of the drug–DNA complex, as compared to the structure of the DNR–CGTACG complex. The
cross-linking reaction is sequence specific in that only a DNA
sequence like 5'-GCG has the proper drug-binding conformation
to place the N-2 amino group of the guanine in the triplet sequence near the N-3' of DNR.

More recently, we have shown that DNR and DOX can also
be cross-linked by HCHO to araC-containing hexamers
CG(araC)GCG and CA(araC)GTG, respectively.[8] These two ad-

ducts provide useful information regarding the influence of another anticancer drug, araC, on the conformation of the anthracycline–DNA complex. The latter complex, DOX–CA(araC)GTG, was crystallized in a different space group (monoclinic $C2$), and its three-dimensional structure was found to be very similar to that of the DNR–CG(araC)GCG complex in the tetragonal $P4_12_12$ space group (Figure 5). This observation reinforces the argument that the crystal lattice forces have only a small influence on the structure of the anthracycline drug–DNA complexes.

The observation that DNR and DOX can be cross-linked to DNA may have significant implication in drug design, since the cross-linking ability of a number of natural antibiotics is well established. For example, several potent antitumor antibiotics act by forming covalent adducts between the drug and DNA. Anthramycin, mitomycin C, saframycin, and ecteinacidins all most likely form covalent adducts with guanine at the N-2 position.[27,28] Interestingly, a highly potent anthracycline antibiotic, barminomycin/SN07, contains an active aldehyde group attached to O-4' of the daunosamine sugar.[29] This aldehyde serves as a cross-linking functional group in ways very similar to the exogenic HCHO discussed above.[30] SN07 has been cross-linked to different DNA polymers [e.g., poly(dG-dC)·poly(dG-dC)], and the resulting drug–DNA adducts appeared to have higher anticancer activities.[29]

Among the new generations of anthracycline drugs, 3'-(4-morpholinyl)-3'-deamino-DOX (MRDox) and its derivative have unusually potent activity when compared with the parent doxorubicin.[31] 3''-Cyanomorpholinodoxorubicin (CN-MRDox) has been suggested to form a covalent cross-link to DNA. Although the exact nature of this adduct is yet to be determined, the mechanism associated with the aldehyde-mediated adduct found in the structure of the DNR–CGCGCG complex may be relevant. We have recently determined the crystal structure of both MRDox and CN–MRDox complexed to CGATCG, which revealed that the C-5 and C-6 positions of the morpholinyl group are in the proximity of bases in the minor groove (unpublished data). The refined structures of the MRDox–CGTACG complex

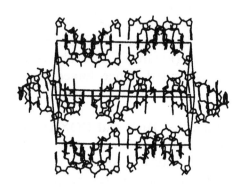

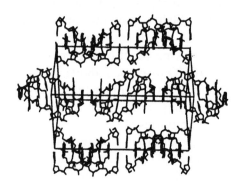

A

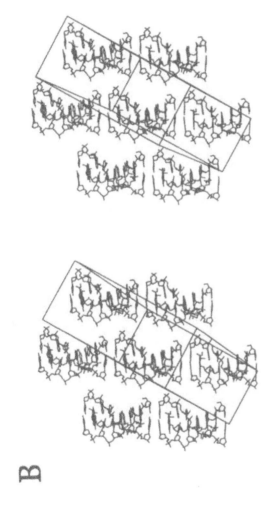

B

Figure 5. Two different crystal lattices for the formaldehyde cross-linked 2:1 complexes of (A) DNR–CG(araC)GCG (space group $P4_12_12$) and (B) DOX–CA(araC)GTG (space group $C2$).

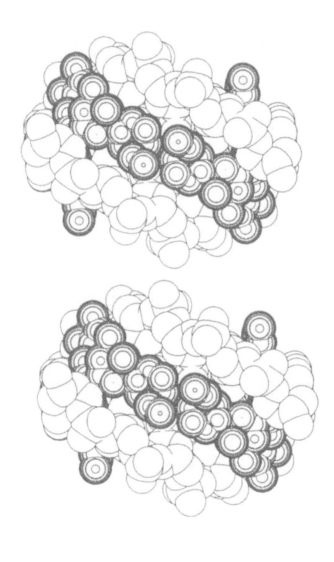

Figure 6. Stereoscopic van der Waals drawing of the 2:1 (*R*)-MMDox–CGTACG complex viewed along the molecular two-fold symmetry. Note that the minor groove is fully occupied.

revealed that two drug molecules bind to the duplex with the aglycones intercalated between the CpG steps with their N-3'-morpholinodaunosamines in the minor groove (Figure 6). The morpholino moiety is flexible and may adopt different conformations dependent upon the sequence context. The O-1'' atoms of the two morpholino groups in the drug–DNA complexes are in van der Waals contact. It is conceivable that for certain DNA sequences the C-5/C-6 position of the morpholino ring may alkylate the N-2 of guanine via the imine intermediate.

The structure of the complex formed from FCE-23672 [a mixture of the (R)- and (S)-isomers of MMDox] has been determined in a new triclinic space group of $P1$.[32] The morpholino rings of the two independent FCE-23672 molecules in the $P1$ crystal lattice adopt different conformations.

VI. NOGALAMYCIN AND ITS DERIVATIVES

Nogalamycin (Ng), because of the bulky sugars attached to both ends of the aglycone, raises an interesting question with respect to the DNA intercalation process. This question has been addressed by several recent studies, including DNase I footprinting experiments[33] and theoretical studies.[34,35]

Earlier NMR studies between Ng and $G_1C_2A_3T_4G_5C_6$ indicated that Ng binds with its aglycone intercalated between the C_2pA_3 and T_4pG_5 steps.[36] Later, more definitive information regarding the detailed interactions was obtained by a series of crystallographic analyses of the complexes between Ng and two modified CGTACG hexamers.[37–39] The X-ray structural analyses (Figure 7) showed that the two Ng molecules are intercalated between the CpG steps at both ends of a distorted B-DNA double helix. The elongated aglycone chromophore (rings A–D) penetrates the DNA double helix such that it is almost perpendicular to the C-1'–C-1' vectors of the two G·C base pairs above and below the intercalator. The drug spans the two grooves of the helix, with the nogalose and the aminoglucose occupying the minor and major grooves, respectively.

Our earlier NMR study of the Ng–CGTACG complex showed that the 2:1 complex in solution is qualitatively similar to the

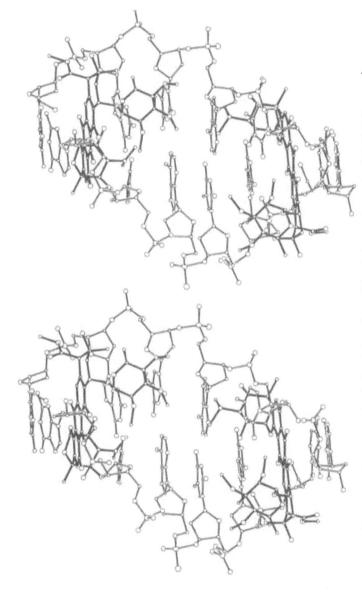

Figure 7. Stereoscopic skeletal drawing of the 2:1 nogalamycin–CGTACG complex.

crystal structure and is very stable, as is evident in the temperature-dependent profiles of the ^{1}H one-dimensional NMR spectra.[40] The exchangeable spectra in H_2O did not reveal the G_6-imino proton, indicating that the terminal base pair is still fraying despite the high stability. The intercalator ring adjacent to the C_1–G_6 base pair apparently did not make the base pair more strongly hydrogen bonded. In contrast, several other exchangeable protons, including the phenolic hydroxyls (O-8 and O-9) of the aglycon ring, are clearly visible. The NOE cross-peaks involving those exchangeable protons are entirely consistent with the refined model to be discussed.

We have undertaken the structural refinement of the symmetric 2:1 Ng–CGTACG complex by NMR.[12] The R factor for the refined Ng–CGTACG model using two-dimensional NOE data is 16.1%, indicating a good agreement between the experimental and simulated NOE data (Figure 3). A conspicuous feature of the Ng–DNA structure is the large buckle between the C_5 and G_6 bases in both the crystal and the solution structures. The refined NMR model is slightly different from the crystal structure in that the upper and lower halves of the solution structure seem to have bent away from each other by about 22°. This difference in the conformation of the solution structure from the crystal structure may be due to the absence of the crystal packing. NMR studies also showed that mobility (the α/β equilibrium) exists in the A ring of Ng, as is evident from the broad resonances associated with the C-13 methyl group and the C-8 (both axial and equatorial) protons. The refined NMR structure showed that the α conformer predominates, with the carbomethoxy group on the C-10 position remaining in the axial position.

In the crystal structure, we detected a smooth bending of the aglycone ring so that the nogalose and the aminoglucose are brought closer together.[37,38] Our NMR refinement resulted in a slightly different conformation of the bound Ng for the solution and crystal structures. They differ primarily in the orientation of nogalose, owing to the altered pucker of ring A. The refined structure of the disnogalamycin–CGTACG complex is generally very similar to the Ng–CGTACG complex. We conclude that disnogalamycin binds to DNA in a manner similar to that of Ng, except that disnogalamycin has a higher mobility in the intercalation cavity.

These different Ng–DNA complexes in solution, including the recent 2:1 Ng–d(GCATGC)$_2$ complex,[36] the 2:1 complex of Ng–d(AGCATGCT)$_2$,[41] the 1:1 complex of Ng–d(GACGTC)$_2$,[42] and Ng–d(GCGT)·d(ACGC)[43] have provided us with an opportunity to examine the sequence specificity of Ng. There are several key hydrogen bonds between the drug and DNA that determine the G·C sequence specificity. We suggest that Ng favors 5′-NpG or 5′-CpN sequences. More specifically, the aglycone chromophore prefers to intercalate at the 5′-side of a guanine (between NpG), or at the 3′-side of a cytosine (between CpN), with the sugars pointing toward the G·C base pair.[38] Fox and Alma showed finer gradations of sequence preference of Ng, even for CpG, depending on the context of the surrounding sequences.[33] The exact reason for these observations requires further structural studies.

VII. ACLACINOMYCIN AND ITS DERIVATIVES

Aclacinomycins A and B (abbreviated AclaA and AclaB, respectively; Figure 1) are other examples of the anthracycline anticancer antibiotics.[44] AclaA has been used in combination with DOX to enhance the effectiveness of DOX on cancer cells that have acquired drug resistance.[45,46] AclaA has an antagonistic effect on DNR-stimulated cleavage of DNA by topoisomerase II.[47] These two new anthracyclines contain a trisaccharide moiety attached to the C-7 position of the A ring of the aglycone chromophore alkavinone (AKN). Their aglycone rings are similar to those of DNR and Ng in that the alkavinone has three unsubstituted C–H positions in ring D and an O-9 hydroxyl group in ring A, as in DNR, but they also have an open H-11 position in ring B and a carbomethoxy group at the C-10 position, as in Ng.

Even though aclacinomycin (Acla) is expected to intercalate DNA, the molecular details of its binding mode (e.g., sequence specificity) remain largely unclear, in contrast to DNR[13] and

Ng.[37] We have obtained crystals of aclacinomycin complexed to DNA, but the quality of these crystals was poor. To obtain the structural information, we have used NMR to study the interaction of AclaA with CGTACG.[48] The titration experiment showed that a stable and symmetric 2:1 complex could be formed. All nonexchangeable resonances of the free DNA and free drugs (AclaA/B) have been unambiguously assigned by 2D-NOESY. The temperature-dependent profiles of the one-dimensional ^{1}H-NMR spectra of the 2:1 AclaA–CGTACG complexes showed small variations in the chemical shifts when the temperature was raised from 5 to 55 °C. The resonances are broad for spectra recorded below 25 °C, suggesting that the binding dynamic of the drug is in the intermediate to slow exchange rate range.

Analysis of the 2D-NOESY spectra indicated that the 2:1 complex has its intercalation sites at the C_1pG_2 and C_5pG_6 steps, with the AKN chromophore oriented such that the ring edge containing H-11 faces toward the backbone of C_1pG_2. A collection of the ten refined structures of the 2:1 AclaB–CGTACG complex obtained from the final ten SPEDREF-SA refinement cycles is shown in Figure 8. The AKN aglycone is intercalated in the CpG step with a small root mean square deviation, indicating a well-defined binding conformation. The cinerulose B has a large deviation, which is most likely due to the sparse number of NOE cross-peaks between protons from cinerulose B and other protons. A conspicuous feature of the refined model is the large buckle of the two GC base pairs (C_1:G_{12} and G_2:C_{11}) that wrap around the aglycone, similar to that of other anthracycline–DNA complexes. There are several possible hydrogen bonds between AclaB and DNA. The trisaccharide lies in the minor groove. The first sugar attached to the chromophore of AKN, the rhodosamine (RN) ring, covers the G_2:C_{11} base pair in the minor groove. The *N*-methyl groups (NMe$_2$ = 3') are proximal to the A_{10}-H-2, -H-1', and -H-4' and C_{11}-H-1', -H-4', -H-5', and -H-5'' protons, consistent with the observed NOEs. Similarly, the RN-H-3' proton has a significant NOE on the A_{10}-H-2 proton.

The DNA duplex is kinked (20°) at the T_3pA_4 step, which may be due to the close contacts between the deoxyfucose sugar

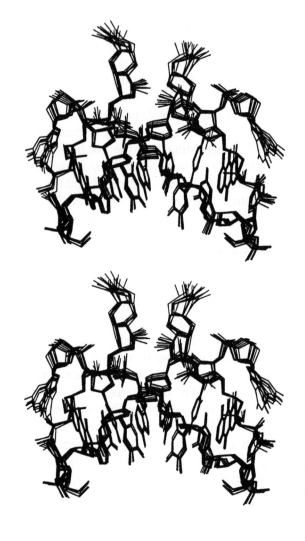

Figure 8. A collection of ten refined structures of the 2:1 complex of aclacinomycin B with d(CGTACG) from use of the SPEDREF procedure (incorporating simulated annealing).

of AclaB and the A_4 deoxyribose of DNA that force the A_4 nucleotide to open up into the minor groove side. Since there are two AclaB molecules in the symmetric complex, the interactions resulting from both AclaB molecules may reinforce each other's role in the deformation of the helix. Thus, the kink in DNA is induced by the positioning of the AKN aglycon in the intercalation cavity, which results in a clash between the deoxyfucose and the A_4 sugar.

Figure 9 compares the intercalation geometries of DOX, Ng, and AclaB in the CpG step. In all three complexes, the aglycone is intercalated in the CpG step, but with a different position and orientation of the elongated ring relative to the surrounding base pairs. DOX has its D ring protruding into the major groove side. Ng, in contrast, has its D ring stacked between the two neighboring base pairs. Since ring A of AclaB has a carbomethoxy group at the C-10 position, the AKN chromophore could not be in a position like that of the DOX complex. An additional carbomethoxy group at the C-10 position in the DNR complex would create a severe clash between the carbomethoxy group and the $C_1:G_{12}$ base pair. Therefore, the 2:1 AclaB–CGTACG complex has its aglycon sliding toward the minor groove (by 1.2 Å) relative to that in the DNR/DOX complex, but maintaining its orientation of the long vector of the aglycone (i.e., no twisting) in the intercalation cavity. This creates a crowding between the drug sugar (dF ring) and the A_4 deoxyribose, which is responsible for the DNA kinking at the TpA step. We also noted that the conformation of *free* AclaB is similar to that of the *bound* form.[48] This observation of a rigid saccharide conformation seems to be quite common, as revealed in the structures of the oligosaccharide antibiotics chromomycin $A_3$49,50 and calicheamicin γ.[51]

We studied the dynamics of the drug binding to DNA using a solution containing 0.5:1 and 1:1 mixtures of AclaA and CGTACG, which shows more resonances than does the 2:1 complex. This is likely to be the result of a mixture of varying population of the three molecular species (free DNA, 1:1 complex, and 2:1 complex), based on the observation that the spectrum is nearly a direct superimposition of the spectra of the

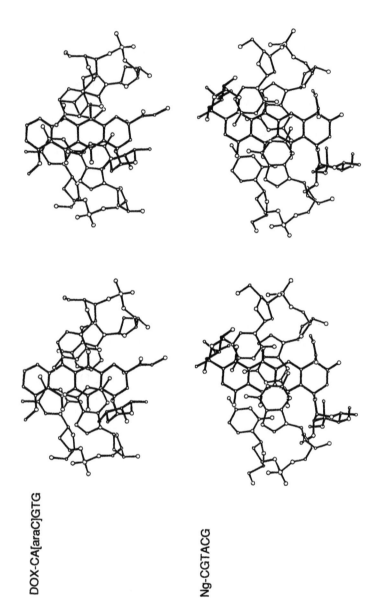

DOX-CA[araC]GTG

Ng-CGTACG

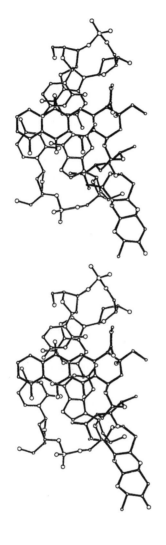

Acla-CGTACG

Figure 9. A view looking perpendicular to the aglycone of three anthracycline drugs in their DNA complexes. The local surroundings of the intercalated drug plus two base pairs (C_1pG_2 and C_5pG_6) of the hexamer helix are shown.

three species. Therefore *the preferred intercalation site of AclaA is CpG*. Further support comes from the 2D-NOESY spectrum of the 1:1 AclaA–CGTACG mixture at 30 °C, which showed numerous exchange peaks. Such an example was also observed in the 1:1 mixture of the Ng–CGTACG complex.[12,40]

The position of the drug binding in the Acla complex is opposite to the echinomycin case (i.e., the two adjacent binding sites are slightly antagonistic to each other). When two AclaB molecules bind simultaneously to the hexamer DNA duplex, the two trisaccharide tails push against the sugar–phosphate backbones, thereby possibly destabilizing the complex owing to their crowding interactions in the minor groove. There may be an exchange between the free and the bound Acla, whose rate is sufficiently slow to allow the exchange cross-peaks to be detected. Indeed, there are several exchange NOE cross-peaks visible in the 2D-NOESY spectrum owing to the slight excess in drug concentration.

VIII. RATIONAL DESIGN OF ANTICANCER DRUGS

An interesting example of a cross-linked drug–DNA complex as cytotoxic agent is the SN07–DNA complex mentioned in Section V.[29] When SN07 is complexed with poly(dG–dC), the complex has significantly higher antitumor activity than SN07 alone (almost 10-fold higher). No such effect was observed when poly(dI–dC) was used, suggesting that the N-2 of guanine is involved in the activity. This led us to believe that a DNA–DNR–HCHO cross-linked adduct may have a similar effect and may be used as a new anticancer agent.

The use of DNA as an enhancing element of cytotoxicity was explored earlier by Trouet and Jolles,[52] who showed that the DNR–DNA noncovalent complex enhanced chemotherapeutic activities on mice. One premise for such an approach is that the DNA–DNR complex may have a lower cardiotoxicity. However, a noncovalent drug–DNA complex tends to dissociate in solution. Therefore, the devising of a method for keeping the drug in the DNA lattice may be useful. Covalent cross-linking, such as the DNA–DNR–HCHO adduct, may be one effective means. DNA, in this case, can act as a carrier for DNR molecules with latent alkylating (aldehyde) function. Since DNR is intercalated and cross-linked by formaldehyde to DNA, the DNR is not free to dissociate. Cells may respond to the cross-linked drug differently. However, in the adduct, DNR molecules can be released from the DNA lattice, because the link between DNR and DNA is an aminal group. Once the cross-linking reaction is reversed, the released drugs should have a cytotoxic effect. Thus, the adduct could have a prolonged cytotoxic effect because of the relatively stable covalent linkage. To test this, we have carried out a number of biological studies. The results are summarized below and have been described in more detail elsewhere.[53]

The adduct was prepared by incubation of the DNR, DNA [either poly(dG–dC) or oligonucleotide $(GC)_{20}$], and HCHO together. The cross-linking density is one DNR per five base pairs, slightly lower than the maximum loading density of one DNR per three base pairs. It should be noted that any anthracycline antibiotic may be used for this purpose, as long as there is an N-3' amino group in the molecule. Therefore, DOX, idarubicin, or other newer generation drugs may be incorporated in a similar manner.

We found that mouse leukemia L1210 cells were sensitive to the purified cross-linked adduct. At high drug doses (e.g., > 1 µg/mL), L1210 cells shrank to about one-fifth the size of the control cells, as observed by microscope, and died by the end of 48 h. At low drug doses, the DNA–DNR–HCHO-treated cells grew at least three times larger than control cells and appeared to be viable until day 5. At a concentration of 64 ng/mL of *free* DNR, most of the treated cells were destroyed at day 3. In contrast, the DNA–DNR–HCHO-treated cells displayed a less rigid appearance. Starting at day 4, some cells showed fragmented morphology and were not viable. The cytosol of the remaining cells was filled with many different-sized vacuoles, which made the cells lumpy and more transparent. These observations suggested that the cross-linked adduct indeed has a significant cytotoxic effect on the L1210 cancer cells.

The IC_{50} cytotoxicity assay of DNA–DNR–HCHO showed that no cells survived at concentrations of 1 µg/mL or higher of cross-linked, non-cross-linked, and free DNR. The estimated IC_{50} was about 10 ng/mL (effective DNR concentration) for cross-linked adducts, about 5 ng/mL (effective DNR concentration) for non-cross-linked adducts, and 5 ng/mL for DNR alone. The relative values among the three compounds suggest that the cross-linked adduct is still active. In comparison, the IC_{50} of DOX was reported to be 20 ng/mL, as measured by the inhibition of growth and macromolecular synthesis.

On the basis of these preliminary data, we suggest that by use of this unique approach the adduct may hold promise as a new chemotherapeutic probe. Our immunological experiments on mice revealed no specific immune response against the DNR–DNA adduct antigens. In fact, it appeared that, despite the distorted DNA conformation, the adduct was still nonimmunogenic. This is not highly surprising as it is well known that double-helical DNA is extremely nonimmunogenic. In contrast, the antibodies against the Z-DNA conformation of poly(dG–dC) have been readily produced.[54] Thus, our results suggest that an undesirable immunological reaction is unlikely.

Another example of an unexpected dividend derives from our structural results of the morpholinodoxorubicins complexed to DNA hexamers, which provide valuable information for the design of new synthetic compounds. We note that the two O-1″ atoms in the 2:1 complex are in van der Waals contacts (Figure 6). It should be possible to substitute the morpholine with either a piperazine or a piperidine and then link two piperazine- or piperidine-modified doxorubicins at their equivalent 1″ positions to make a bisintercalating DOX. This new type of bis-doxorubicin has several attractive features. It is likely to act as a true bisintercalator, since the two doxorubicins are linked by a relatively rigid tethering linker, that is, a pair of piperazine- or piperidine-modified daunosamines. A rigid tether is necessary for efficient bisintercalation, as demonstrated in the case of bis-intercalator antibiotics triostin A[55] and synthetic ditercalinium.[56] These new bis-doxorubicins may also have a higher DNA affinity, as well as a heretofore unseen binding specificity for a longer sequence such as 5'-CGATCG. As they are structurally quite different from DOX itself, the DOX-resistant cancer cells may be sensitive to them. Syntheses of such compounds are now in progress.

IX. SUMMARY

Crystal structure analysis of several complexes between important anthracycline drugs and DNA oligonucleotides affords us valuable information for understanding the role of various functional groups of the drug molecules. Moreover, the excellent agreement between the crystal and solution structures shown in the analyses of the Ng–CGTACG complex showed that a quantitative treatment of NOE data is an effective alternate way of producing a reliable three-dimensional structure of drug–DNA complexes. Aclacinomycins A and B, complexed to DNA by the same procedure, provided us with a structure that is not yet available from X-ray crystallography. Similar analyses of other drugs, such as arugomycin,[57] will be useful in our understanding of the function of anthracycline antibiotics in general.

The structural results also offer some clues for the design of new and unique classes of potential drugs. For example, our preliminary data on the nature of the drug–DNA cross-link suggest that such a complex has several potential beneficial properties as an anticancer drug:

1. The conjugate can be prepared readily and efficiently with very high loading of the drug [to a maximum of one DOX per three base pairs for poly(dG–dC)].
2. The conjugate may be long-lived in the blood stream, serving as a slow-releasing drug reservoir.
3. The cytotoxic agent is hidden in the DNA lattice, avoiding the attack by enzyme that causes the formation of the free radical species of DOX. This may reduce the cardiotoxicity side effect.
4. If the conjugate can be picked up by the cancer cells directly, it may be effective against resistant cells.
5. The conjugate should be resistant to nuclease, since the DNA conformation is severely distorted by the intercalated drug.

6. A sequence-specific DNA–DOX adduct may be prepared, as long as some guanine nucleotides are present.

We believe that continuing structural studies on many drugs bound to their receptors will be extremely fruitful in guiding the design of new drugs.[58]

ACKNOWLEDGMENTS

This work was supported by grants from the National Institutes of Health (GM-41612 and CA-52506) and the American Cancer Society (DHP-114). I thank Y.–g. Gao, H. Robinson, D. Yang, and J. Y.–T. Wang for their contributions.

REFERENCES

1. *Anthracycline and Anthacenedione-based Anticancer Agents;* Lown, J. W., Ed.; Elsevier: New York, 1988.
2. Denny, W. A. *Anti-Cancer Drug Design* **1989**, *4*, 241–263.
3. *Molecular Basis of Specificity in Nucleic Acid–Drug Interactions;* Pullman, B.; Jortner, J., Eds.; Kluwer Academic Publishers: Dordrecht, 1990.
4. Wang, A. H.-J. *Current Opin. Struct. Biol.* **1992**, *2*, 361–368.
5. Drenth, J. *Principles of Protein Crystallography;* Springer-Verlag: New York, 1994.
6. McPherson, A. *Preparation & Analysis of Protein Crystals;* Wiley: New York, 1982.
7. Brunger, A. T. *X-PLOR,* Version 3.1; The Howard Hughes Medical Institute & Department of Molecular Biophysics & Biochemistry, Yale University, 260 Whitney Avenue, P. O. Box 6666, New Haven, CT 06511, 1992.
8. Zhang, H.; Gao, Y.-G.; van der Marel, G.A.; van Boom, J. H.; Wang, A. H.-J. *J. Biol. Chem.* **1993**, *268*, 10095–10101.
9. Wang, A. H.-J.; Quigley, G. J.; Kolpak, F. J.; Crawford, J. L.; van Boom, J. H.; van der Marel, G. A.; Rich, A. *Nature* **1979**, *282*, 680–686.
10. Robinson, H.; Wang, A. H.-J. *Biochemistry* **1992**, *31*, 3524–3533.
11. Keepers, J. W.; James, T. *J. Mag. Reson.* **1984**, *57*, 404–426.
12. Robinson, H.; Yang, D.; Wang, A. H.-J. *Gene* **1994**, *149*, 179–188.
13. Wang, A. H.-J.; Ughetto, G.; Quigley, G. J.; Rich, A. *Biochemistry* **1987**, *26*, 1152–1163.
14. Gao, Y.-G.; Sriram, M.; Wang, A. H.-J. *Nucleic Acids Res.* **1993**, *17*, 4093–4101.
15. Wang, A. H.-J.; Gao, Y.-G.; Liaw, Y.-C.; Li, Y. K. *Biochemistry* **1991**, *30*, 3812–3815.

16. Gao, Y.-G.; Liaw, Y.-C.; Li, Y. K.; van der Marel, G. A.; van Boom, J. H.; Wang, A. H.-J. *Proc. Natl. Acad. Sci. U.S.A.* **1991**, *88*, 4845–4849.

17. Frederick, C. A.; Williams, L. D.; Ughetto, G.; van der Marel, G. A.; van Boom, J. H.; Quigley, G. J.; Rich, A.; Wang, A. H.-J. *Biochemistry* **1990**, *29*, 2538–2549.

18. Moore, M. H.; Hunter, W. N.; Langlois D'Estaintot, B.; Kennard, O. *J. Mol. Biol.* **1989**, *206*, 693–705.

19. Chaires, J. B. In *Advances in DNA Sequence Specific Agents.* Vol. 1.; Hurley, L. H., Ed.; JAI Press Inc.: Greenwich, 1992, pp. 3–23.

20. Chen, K. X.; Gresh, N.; Pullman, B. *J. Biomolec. Struct. Dynam.* **1985**, *3*, 445–466.

21. Nunn, C. M.; Meervelt, L. V., Zhang, S.; Moore, M. H.; Kennard, O. *J. Mol. Biol.* **1991**, *222*, 167–177.

22. Leonard, G. A.; Hambley, T. W.; McAuley-Hecht, K.; Brown, T.; Hunter, W. N. *Acta Cryst.* **1993**, *D49*, 458–467.

23. Lipscomb, L. A.; Peek, M. E.; Zhou, F. X., Bertrand, J. A.; VanDerveer, D.; Williams, L. D. *Biochemistry* **1994**, *33*, 3649–3659.

24. Williams, L. D.; Egli, M.; Ughetto, G.; van der Marel, G. A.; van Boom, J. H.; Quigley, G. J.; Wang, A. H.-J.; Rich, A.; Frederick, C. A. *J. Mol. Biol.* **1990**, *215*, 313–320.

25. Gao, Y.-G.; Wang, A. H.-J. *Anti-Cancer Drug Des.* **1991**, *6*, 137–149.

26. Williams, L. D.; Ughetto, G.; Frederick, C. A.; Rich, A. *Nucleic Acids Res.* **1990**, *18*, 5533–5541.

27. Warpehoski, M. A.; Hurley, L. H. *Chem. Res. Toxicol.* **1988** , *1*, 315–333.

28. Guan, Y.; Sakai, R.; Rinehart, K. L.; Wang, A. H.-J. *J. Biomolec. Struct. Dynam.* **1993**, *10*, 793–818.

29. Kimura, K.; Takahashi, H.; Takaoka, H.; Miyata, N.; Nakayama, S.; Miyata, N.; Kawanishi, H. *Agri. Biol. Chem.* **1990**, *54*, 1645–1650.

30. Ye, X.; Kimura, K.; Patel, D. *J. Am. Chem. Soc.* **1993**, *115*, 9325–9326.

31. Acton, E. M.; Tong, G.; Mosher, C.; Wolgemuth, R. *J. Med. Chem.* **1984**, *27*, 638–645.

32. Cirilli, M.; Bachechi, F.; Ughetto, G.; Colonna, F. P.; Capobianco, M. L. *J. Mol. Biol.* **1993**, *230*, 878–889.

33. Fox, K. R.; Alam, Z. *Eur. J. Biochem.* **1992**, *209*, 31–36.

34. Brown, J. R.; Neidle, S. In *Anthracycline and Anthracenedione-Based Anticancer Agents;* Lown, J. W., Ed.; Elsevier: New York, 1988.

35. Trinquier, G.; Chen, K. X.; Gresh, N. *Biopolymers* **1988**, *27*, 1491–1517.

36. Searle, M. S.; Hall, J. G.; Denny, W. A.; Wakelin, L. P. G. *Biochemistry* **1988**, *27*, 4340–4349.

37. Liaw, Y. C.; Gao, Y. G.; Robinson, H.; van der Marel, G. A.; van Boom, J. H.; Wang, A. H.-J. *Biochemistry* **1989**, *28*, 9913–9918.

38. Gao, Y. G.; Liaw, Y. C.; Robinson, H.; Wang, A. H.-J. *Biochemistry* **1990**, *29*, 10307–10316.

39. Egli, M.; Williams, L. D.; Frederick, C. A.; Rich, A. *Biochemistry* **1991**, *30*, 1364–1372.

40. Robinson, H.; Liaw, Y. C.; van der Marel, G. A.; van Boom, J. H.; Wang, A. H.-J. *Nucleic Acids Res.* **1990**, *18*, 4851–4858.
41. Zhang, X.; Patel, D. J. *Biochemistry* **1990**, *29*, 9451–9466.
42. Searle, M.; Bicknell, W. *Eur. J. Biochem.* **1992**, *205*, 45–58.
43. van Houte, L. P. A.; van Garderen, C. J.; Patel, D. J. *Biochemistry* **1993**, *32*, 1667–1674.
44. Oki, T.; Takeuchi, T.; Oka, S.; Umezawa, H. *Recent Results Cancer Res.* **1981**, *76*, 21–40.
45. Millot, J.-M.; Rasoanaivo, T. D.; Morjani, H.; Manfait, M. *Brit. J. Cancer* **1989**, *60*, 678–684.
46. Tapiero, H.; Boule, D.; Trincal, G.; Fourcade, A.; Lampidis, T. J. *Leu. Res.* **1988**, *12*, 411–418.
47. Jensen, P. B.; Jensen, P. S.; Demant, E. F. J.; Friche, E.; Sorensen, B. S.; Sehested, M.; Wasserman, K.; Vindelov, L.; Westergaard, O.; Hansen, H. H. *Cancer Res.* **1991**, *51*, 5093–5099.
48. Yang, D.; Wang, A. H.-J. *Biochemistry* **1994**, *34*, 9451–9466.
49. Silva, D. J.; Kahne, D. E. *J. Am. Chem. Soc.* **1993**, *115*, 7962–7970.
50. Gao, X.; Mirau, P.; Patel, D. *J. Mol. Biol.* **1992**, *223*, 259–279.
51. Walker, S.; Murnick, J.; Kahne, D. E. *J. Am. Chem. Soc.* **1993**, *115*, 7950–7961.
52. Trouet, A.; Jolles, G. *Seminar in Oncology* **1984**, *11*, 64–72.
53. Wang, J. Y.-T.; Cho, M.; Wang, A. H.-J. In *Anthracyclines;* Priebe, W., Ed.; American Chemical Society: Washington, D.C., 1994.
54. Stollar, B. D. *CRC Rev. Biochem.* **1986**, *20*, 1–36.
55. Wang, A. H.-J.; Ughetto, G.; Quigley, G. J; Hakoshima, T.; van der Marel, G. A.; van Boom, J. H.; Rich, A. *Science* **1984**, *225*, 1115–1121.
56. Delepirre M.; Maroun R.; Garbay-Jauregilberry, C.; Igolen, J.; Roques, B. P. *J. Mol. Biol.* **1989**, *210*, 211–228.
57. Searle, M.; Bicknell, W.; Wakelin, L. P. G.; William, A. D. *Nucleic Acids Res.* **1991**, *19*, 2897–2906.
58. Gao, Y.-g.; Wang, A. H.-J. *J. Biomolec. Struct. Dynam.* **1995**, *13*, 103–117.

TRANSCRIPTIONAL ASSAY FOR PROBING MOLECULAR ASPECTS OF DRUG–DNA INTERACTIONS

Don R. Phillips

Advances in DNA Sequence Specific Agents
Volume 2, pages 101–137.
Copyright © 1996 by JAI Press Inc.
All rights of reproduction in any form reserved.
ISBN: 1-55938-166-3

ABSTRACT

An in vitro transcription assay of drug–DNA interactions has been developed largely on the basis of the stable *lac*-UV5-initiated transcription complex. This system comprises a synchronized population of radiolabeled nascent RNA ten nucleotides long. Reaction of drugs with this initiated transcription complex, followed by elongation of the nascent RNA by *E. coli* RNA polymerase, reveals blockages at drug sites. From these blockages it is possible to obtain four features of the drug–DNA interaction: the sequence of preferred drug binding sites; the relative drug occupancy at each site; the drug dissociation rate at each site; and the probability of drug-induced termination of transcription at each site. The unidirectional transcription assay has also been extended to a two-promoter, counter-directed system, which yields a bidirectional transcriptional footprint of drug sites. The transcription assay has recently been applied to in vitro eukaryotic systems in which drug-induced blockages of RNA polymerase II reveal drug binding sites and the capacity of elongation factors to influence the effect of drugs at these sites.

I. INTRODUCTION

Of the fifty or so registered nonsteroid anticancer agents,[1-3] approximately 50% are known to interact with DNA. Although their mode of interaction with DNA is varied, encompassing intercalation (actinomycin D, adriamycin, amsacrine, daunomycin, mitoxantrone), binding in the minor groove (distamycin),

strand scission (bleomycin), alkylation (nitrogen mustard, mel-
phalan), and interstrand cross-linking (carboplatin, cisplatin, mitomycin C), a fundamental characteristic of these drugs is that
they all exhibit sequence selective binding to DNA. There have
therefore been extensive efforts over the last two decades to
gain improved understanding of the nature of this sequence
specificity. The expectation is that a precise knowledge of the
factors affecting this specificity will enhance rational design
and development of improved derivatives of these drugs.

Initial attempts to elucidate the sequence specificity of these
drugs relied largely upon physicochemical techniques. The advent of the "molecular biology era" provided a range of alternative procedures to probe the sequence specificity of
drug–DNA interactions.[4–7] The most widely employed procedure
has been that of footprinting, using either DNase I,[8,9] MPE-
Fe(II),[8] or Fe-EDTA,[10] and these approaches have been reviewed
in detail.[5,6,11,12]

Although the DNase I and MPE-Fe(II) footprinting procedures
have been extremely successful in yielding the sequence specificity of a range of DNA-binding ligands, there has also been
an increasing desire to obtain this information under a more
physiological situation and in a more quantitative manner. In
principle, this could be achieved by the monitoring of blockages
to the processive movement of either DNA polymerase or RNA
polymerase along the DNA template. DNA polymerase has been
used successfully to probe the sequence specific localization of
many drugs in vitro,[13,14] and this has also been achieved in vivo
by use of the highly repeated alpha DNA fragment.[15] However,
the nature of the initiation process is such that additional quantitation of drug occupancy and drug kinetics is not readily accomplished.

The utility of RNA polymerase as a probe of drug sites on
DNA was originally demonstrated with RNA primers of varying
lengths[16]; subsequent extension of the RNA in the presence of
drug was therefore not suitable for quantitative analysis. Reported here is a procedure developed to synchronize the initiation of transcription so that all promoter-containing DNA
fragments are transcribed to the same length; subsequent elon-

gation of the transcription process therefore enables all RNA polymerases to move simultaneously toward drug blockage sites. This in vitro transcription assay therefore offers the desirable feature that many aspects of the drug–DNA interaction can be quantitated: not only is the sequence specificity revealed, but also the relative drug occupancy at each site, the rate of dissociation of drug from each individual site, and the probability of termination at each individual drug site. Importantly, all of this quantitative information is obtained under conditions of active transcription of the DNA and therefore provides some insight into the processes that may occur in a cellular environment.

II. THE TRANSCRIPTION ASSAY

A. Overview of the Procedure

The transcription assay relies upon there being a homogeneous population of DNA fragments (containing a promoter), all in the form of an initiated transcription complex in which the nascent RNA is of a constant length and the entire population of transcription complexes is synchronized to begin elongation from a common site. Exposure of this complex to a drug, followed by subsequent elongation of the nascent RNA, leads to the processive movement of the RNA polymerase along the DNA template until it is retarded by the presence of a drug. The length of the RNA (the blocked transcript) therefore reveals the location of the drug site (Figure 1). The relative amount of blocked transcript is related to the relative drug occupancy at that site, and the rate of disappearance of the blockage is related to the rate of dissociation of drug from that site.

B. History of Development

There have been four stages of development of this technique. The initial stage involved a preliminary demonstration of the viability of the technique using the *E. coli lac* UV5 promoter system.[17] This system was chosen because the molecular details

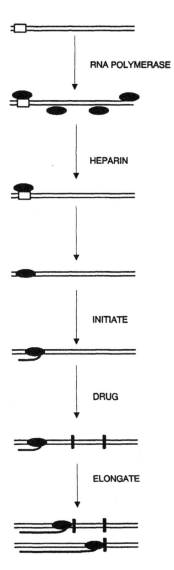

Figure 1. Overview of the transcription assay. The major steps are (1) formation of a synchronized initiated transcription complex (see Figure 2 for details), (2) reaction of the initiated transcription complex with drug, and (3) elongation of the transcription complex to yield drug-induced blocked transcripts.

of the initiation process were largely resolved[18] and the characteristics of the promoter–RNA polymerase interaction were well described.[19] The first demonstration of the utility of this system was a study of actinomycin D, a model drug with well-characterized sequence specificity. Only one blockage site was detected in the 67 nucleotides of DNA transcribed in that system.[17] In the second phase of development of this technique, the length of DNA transcribed from the UV5 promoter was extended to several hundred nucleotides in order to probe a greater range of unique sequences, and the elongation conditions were optimized in order to minimize natural pausing.[20]

The third phase of development was directed to a more quantitative analysis of the transcriptional blockages, and this was initially accomplished by use of a single statistical correction term to the apparent drug occupancy to correct for decreasing amounts of RNA polymerase reaching downstream sites.[21] Subsequently, a Monte Carlo simulation procedure was established to permit a complete description of multiple drug sites in terms of drug occupancy, drug dissociation kinetics, and probability of termination at each drug site.[22] A simple and rapid compartmental simulation is now available for this purpose and can be carried out in minutes on a desktop personal computer.[23] The final phase of development has resulted in the technique of "bidirectional transcription footprinting," in which two promoters are in a counter-directed orientation. This procedure yields transcriptional blockages from each promoter on the same fragment of DNA.[24]

The experimental aspects of this technique have recently been outlined in detail,[25] and there are now several reviews of the application of this procedure to the study of drug–DNA interactions.[26,27]

C. Synchronized Initiation

There are several characteristics required of any promoter in this assay: they should preferably not require activating elements; they should exhibit good fidelity at the start of transcription (or be able to be forced to a high fidelity at the start

Table 1. Promoter Systems That Yield Synchronized Initiated
Transcription Complexes

Promoter	Initiation Dinucleotide	Nucleotide Absent During Initiation	Length of Initiated Transcript (nuc)	Reference
UV5	GA	CTP	10	17,20
N25	AU	CTP	29	24
Tet[R]	AG	GTP	11	28
SP6	AG	GTP	9	21
T3	AG	CTP	12	21
T7	AG	UTP	13	21
λP_L	AU	UTP	15	28

site); the DNA sequence must be such that a stable initiated transcription complex can form in the absence of one or more nucleoside triphosphates in the initiation mixture; and the initiated transcription complex should be stable, with a half-life preferably of several hours. Several promoters satisfy these criteria to varying degrees; these are summarized in Table 1.[17,20,21,24,28] Because the UV5 promoter has been extensively characterized, it was used for all initial work in this assay and subsequent studies.

The most critical aspect of the assay is formation of an initiation complex in which all of the nascent RNA is of exactly the same length and the entire population of transcripts is poised to continue transcription from a common site. With the UV5 promoter, this synchronization has been achieved by use of high levels of the dinucleotide GpA, which forces transcription to begin selectively from one start site (complementary to the −1 and +1 positions) and allows transcription to proceed up to a nucleotide (CTP) that is absent from the transcription mixture. This yields RNA comprising mainly a 10-mer (Figure 2), together with small amounts of 17- and 23-mers.[20] The reason for some read-through past the missing C residue at position +10 is thought to be because CTP is present in trace amounts,

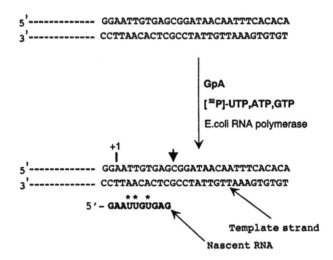

Figure 2. Synchronized initiated transcription complex. Initiation of the *lac* UV5 promoter with GpA, ATP, GTP, and [α-^{32}P]UTP results in a stable transcription complex containing nascent RNA, mainly 10 nucleotides in length, up to C of the nontemplate strand (denoted with an arrow), since CTP is absent from the initiation nucleotide mixture. The nascent RNA begins at the −1 position with G of GpA in the initiation mixture. Radiolabel (^{32}P) is incorporated into the nascent RNA at three sites, denoted with an asterisk.

even in high-purity nucleotides, but it could also be that small amounts of other nucleotides are misincorporated.

It is essential, for the subsequent analysis of lengths of transcripts, that every molecule of RNA polymerase begins transcription at only one site. This does not normally occur under in vitro conditions, with only 59% of transcripts initiating at the +1 position of the UV5 promoter when all four NTPs are present, and with 29%, 7%, and 9% beginning at the −1, +2, and +5 positions, respectively.[29] This lack of homogeneity of the start site was solved by restriction of all NTPs present to 5 μM, and because this level is too low for incorporation of the first nucleotide into the transcription complex (the K_M is higher for the first nucleotide than for all subsequent incorporations), significant initiation does not occur. The presence of a high level of the first nucleotide would then permit the initiation process to commence. In practice, to add further specificity to the process, the first dinucleotide is usually added at high concentration to minimize the possibility of initiation from any other site.

An additional requirement of the assay is that the initiated transcription complex be sufficiently stable to permit the desired reactions with drugs to be performed. The 10-mer (formed by initiation from the −1 and +1 sites with GpA) has a half-life of 23 h,[30] and has proved to be sufficiently stable for all drug–DNA studies to date. If greater stability is required, an 11-mer can be used. It can be formed by initiation from the −2 position with high levels of GpGpA in the initiation mixture.[18] The individual steps in the formation of the initiated complex are

1. Addition of RNA polymerase (approximately 100 nM) to the promoter-containing DNA fragment (approximately 50 nM); 1 min is sufficient to enable the open complex to form, although 15 min at 37 °C is used routinely.
2. Addition of heparin (400 μg/mL, 5 min) to displace RNA polymerase from nonspecific binding sites on the DNA (including the ends of the DNA fragment) and to ensure that only single-copy transcripts result during the elongation phase (i.e., the RNA polymerase is unable to rebind to the promoter).
3. Addition of GpA (200 μM) and 5 μM of ATP, GTP, and [α-^{32}P]UTP. The typical reaction mixture at this stage is 20–100 μL.

These steps are outlined in Figure 1, together with subsequent steps of the transcription assay.

The transcription buffer routinely used comprises 40 mM Tris-HCl (pH 8.0), 100 mM KCl, 3 mM $MgCl_2$, 0.1 mM EDTA, 5 mM DTT, 0.5 mg/mL acetylated BSA (or nuclease-free BSA), and 1 unit/μL RNase inhibitor. The use of RNase inhibitor is optional for short reaction and elongation times but is necessary for reactions in the 2–20 h time range. The $MgCl_2$ concentration is critical to ensure efficient transcription and minimal natural pausing.[31]

D. Equilibration with Drug

The initiated complex is reacted with the drug for the appropriate time to ensure that equilibrium has been attained (typically 5–30 min for reversibly binding drugs). The maximum reaction time for this step is approximately 48 h at 37 °C (given the half-life of the initiated complex of 23 h), but this could be prolonged at lower temperatures. Subsaturating levels of drug are normally used to enable the RNA polymerase to reach a range of different drug sites in the elongation phase. Saturating levels of drug would result in each drug site being fully occupied; the polymerase would therefore be blocked at the first site, and additional sites further downstream could not be probed. Therefore, although the sensitivity of detection of the first drug site would be increased under saturating conditions, the information content would be decreased dramatically.

E. Elongation

Addition of high levels (2.5 mM) of all four nucleotides enables a synchronized and rapid elongation of the initiated com-

---

Figure 3. Transcription blockages induced by nogalamycin. (A) Nogalamycin (6 µM) was reacted for 1 h at 37 °C with a 497-bp DNA fragment containing the initiated UV5 promoter, and elongation was then carried out at 37 °C. Lanes 0.25 to 420 denote the time course of elongation (minutes) following addition of an elongation mixture comprising 2.5 mM of all four nucleotides in 0.4 M KCl. DNA in the control lanes was subjected to the same procedure in the absence of nogalamycin. Sequencing lanes in which elongation of the initiated complex was terminated by the presence of 3'-methoxy ATP and 2'-methoxy CTP are denoted as A and C, respectively. The full-length transcript is denoted as FT. (B) The relative intensity of nogalamycin-induced transcriptional blockages (expressed as relative occupancy of drug at each site) is shown with respect to the sequence of the nontemplate DNA strand. The consensus specificity for 5'-CA sites is indicated by transcriptional blockages immediately preceding all 5'-CA (and 5'-TG) sequences, and these have been highlighted by underlining.

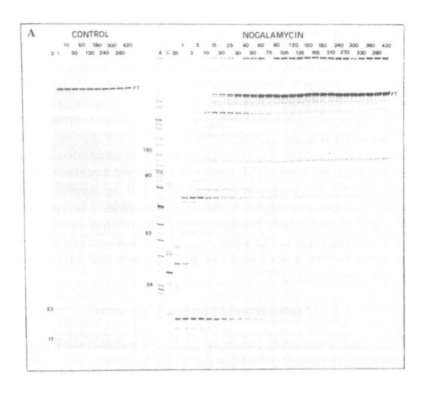

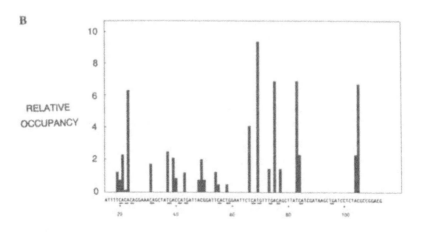

plex up to the first occupied drug site (see Figure 1). Sequencing gels of the labeled RNA therefore reveal a range of drug-induced blocked transcripts at short times after commencement of elongation. An example of blockages induced by nogalamycin is shown in Figure 3. In initial studies of drug-induced pausing, elongation conditions were used that allowed incorporation of ^{32}P-label into the growing RNA chain.[17] Although this procedure yields increasing sensitivity of detection of blockage sites with increasing transcript length, it also introduces additional problems for quantitation. High levels of unlabeled nucleotide are now used in the elongation phase, and further incorporation of label into the growing RNA chain is negligible. The relative intensity of each band therefore reflects the relative amount of RNA of each length. The effect of natural pausing is now also minimized by the use of high salt concentration (0.4 M) during the elongation phase.[20]

F. Quantitation of Blocked Transcripts

The RNA transcripts resulting from elongation up to drug blockage sites are resolved by conventional denaturing, high-resolution sequencing gels run at approximately 60 °C. To increase the number of resolvable bands on an autoradiogram (suitable for quantitative purposes), wedge gels can be used, but usually it is simpler to use a double-loading procedure to increase the resolvable bands from the 20–90 range up to 90–150 (when subjected to twice the normal time of electrophoresis).

Quantitation of the relative amount of each RNA can be performed either by conventional autoradiography and scanning densitometry or by a phosphorimaging process. Autoradiography is usually carried out overnight at room temperature without intensifier screens in order to maximize the resolution of bands. The time of exposure must be modified to ensure that the absorbance of all bands is within the region of photographic linearity; for Kodak XAR X-ray film, linearity is limited to the 0–1 absorbance range.[32] For maximal resolution, the gels are fixed and dried prior to autoradiography. The fixing step in this procedure also serves to remove almost all background on the gels arising from radiolytic degradation products of the labeled nucleotide(s). Amersham β-max X-ray film is routinely used for the final quantitative work because of the low background absorbance and high contrast of this film. Quantitation of band intensity is then performed with a densitometer. Although simple, conventional laser densitometers are adequate, the data obtained are restricted to a spot or narrow slit that covers only part of each band.[22] We have recently shown that the typical reproducibility of these data is 10–30% for repeated rescans of the same autoradiogram, with the larger error being associated with low intensity bands. Significantly better reproducibility is possible if densitometry is performed with a 2D-densitometer capable of quantitating "volume elements". By this means, the absorbance is summed over the entire area of the band to yield the total absorbance of each volume element, which is then independent of band shape.

A vastly superior form of analysis of the sequencing data utilizes the technique of phosphorimaging. This procedure offers three major advantages over conventional densitometry: speed, sensitivity, and dynamic range. The phosphorimager is 10–100 times more sensitive than autoradiography (and therefore produces the same level of band intensity much faster) and has a dynamic range at least 400 times that of photographic film. For these reasons it is the method of choice. Following exposure of the dried gel to a phosphor plate for several hours, the plate is then scanned by the phosphorimaging system, and the absolute intensity, relative intensity, or mole fraction of RNA in each band is then readily determined. An example of the mole fraction of blocked transcripts induced by nogalamycin is shown in Figure 3B.

G. Experimental Precautions

Radiolabeled nucleotides older than two weeks can interfere with the RNA polymerase activity. This problem is more acute with bacteriophage RNA polymerases than with bacterial RNA polymerases. In general, the fresher the radiolabel, the fewer the problems during both initiation and elongation phases.

Purity of the promoter-containing DNA fragment is critical. If a range of sizes of elongation products due to background pausing is observed, this can often be rectified by subjecting the DNA to a clean-up procedure, such as a NENSORB20 nucleic acid cartridge (NEN Research Products, DE).

III. SEQUENCE SPECIFICITY

A fundamental characteristic of any drug–DNA complex is the sequence of the DNA that forms the complexes. One of the major attributes of the transcription assay is the clarity with which drug sites can be resolved, with sequence specificity usually being readily defined to within a base pair. An example of the precision with which preferred drug sites can be defined is shown for nogalamycin (Figure 3), where five of the six dominant sites immediately precede 5'-CA sequences. The assay has now been applied to a series of drugs of clinical or preclinical significance, and these are summarized below.

In the early stages of application of this procedure, it was not known if any transcriptional blockages arose from impairment of the processive capacity of *E. coli* RNA polymerase at sites of depurination or single-strand nicks. It is now clear that *E. coli* RNA polymerase can bypass apurinic sites, and this is accompanied by the incorporation of adenine into the nascent RNA at those sites.[33] Although some termination of transcription has been reported in the region of the apurinic site,[32] there is a slow read-through (half-life of approximately 10 h) past nitrogen-mustard-induced monoadducts, and this is consistent with the rate of depurination at these sites (approximately 9 h).[34] Single-strand nicks do not affect the transcriptional activity of bacteriophage RNA polymerase,[35] and this also appears to be true for *E. coli* RNA polymerase.[36] However, there is complete termination at sites that are apurinic and contain a single-strand nick.[33]

A. Reversible Drug–DNA Interactions

All of the reversibly acting drugs studied by this procedure have been intercalators, and their observed sequence specificity is summarized in Table 2.[4,20,22,24,28,37–45] Blocked transcripts are generally observed up to the base pair that is part of an intercalation site (see, for example, Figure 3), but these are also sometimes seen 1 bp before the intercalation site. That transcription proceeds right up to the drug intercalation site is somewhat surprising, given the structure of the initiated transcription complex, and suggests that the catalytic site is several base pairs back from the physical boundary of the polymerase. This is particularly evident for actinomycin D, where it is known that the pentapeptide chains protrude into the minor groove,[46] yet transcription proceeds right up to the intercalation site. This indicates that *E. coli* RNA polymerase must "track" in the major groove of DNA.

With two drugs, echinomycin and actinomycin D, blockages have been observed downstream from the drug site. In the case of echinomycin this is presumably because it is a bisintercalator, with the "linking unit" protruding into the minor groove. Because of the helicity of the DNA, and the fact that the RNA polymerase tracks in the major groove, the physical blockage therefore occurs only at (or just past) the first intercalation moiety. With actinomycin D, transcriptional blockages were ob-

Table 2. DNA Sequence Specificity of Reversibly Binding Drugs

Drug	Sequence[a]	Reference	Sequence[b]	Reference
Actinomycin D	GC	20	GC	4,40
Adriamycin	TCA	28	—	—
Daunomycin	CA	37,38	(A/T)GC(A/T)CG	41,42
Echinomycin	CG	22	CG	43,44
Mithramycin	GC	24	(G/C)(G/C)	45
Mitoxantrone	CA	39	—	—
Nogalamycin	CA	24	CA	45

Notes: [a]Unidirectional in vitro transcription (*lac* UV5 promoter).
　　　 [b]Other procedures.

served 7–10 nucleotides downstream of the original blockage, following extensive elongation times. This appears to be a special case and has been referred to as "delayed termination". It is thought to arise from secondary structures in the RNA that are transcribed after the drug has dissociated from that site and polymerase has bypassed the original drug site.[20]

In general, intercalating drugs result in transcriptional blockages at a variety of DNA sequences, and it is the common sequence features at these sites that enable a consensus sequence to be identified. For example, with adriamycin at 10 °C, 12 major blockage sites have been identified, 9 of which immediately preceded CpA sequences (suggesting a preference for intercalation of adriamycin between C and A), and of these, 7 contained a thymine 5' to the drug site. The consensus sequence specificity under these conditions is therefore 5'-TCA.[28]

For drugs that interact reversibly with DNA, the residence time is often too short to induce a significant number of transcriptional blockages. The smallest complex half-life that will allow use of this procedure is approximately 10 s at 37 °C. For half-lives that are smaller, it is necessary to slow the drug dissociation rate by a decrease in the temperature of the elongation phase. The lowest temperature that can be employed is approximately 10 °C. Below this temperature, pausing of transcription in the absence of drug becomes significant and precludes any meaningful analysis of drug-induced blockages. Use of reduced temperature has enabled the sequence specificity of rapidly dissociating drugs, such as adriamycin and mitoxantrone, with residence times of approximately 1 s at 37 °C, to be determined.[28,39]

B. DNA Adducts

The sequence specificities of a number of drugs that form adducts have now been established by use of the transcription assay (Table 3).[30,47–56] The experimental procedure for the analysis of these drugs is simpler than for reversibly interacting compounds such as intercalators, where the drug dissociation process necessitates rapid sampling of the transcription complex at early elongation times, and may also necessitate the use of low tem-

Table 3. DNA Sequence Specificity of Drug-Induced Adducts

Drug	Sequence^a	Reference	Sequence^b	Reference
Adriamycin	GC	30,47	—	—
Cisplatin	GG, AG	48	$G_n (n > 2)$	51
Cyanomorpholino-adriamycin	GG, GC	49	GG	49
Nitrogen mustard	G, GG	50	G, G_n	52,53
Mitomycin C	G, CG, GG	47	G, CG, GG	54–56

Notes: ^{a}Unidirectional in vitro transcription (*lac* UV5 promoter).
        ^{b}Other procedures.

peratures. The transcriptional blockages are also much easier to interpret, since the relative intensity of blockages does not generally vary with the extent of elongation time. This permanence of blockage intensities makes it relatively simple to identify major blockage sites (see, for example, Figure 4, where 12 of the 14 high-intensity blockage sites precede 5′-GG sequences on either the template or nontemplate strands). Even so, it is necessary to ensure that a consensus site of interaction emerges from blockages at a multitude of sites. This is especially true if transcription is terminated when the adduct is on the template strand, but not when on the nontemplate strand. Cisplatin and nitrogen mustard, for example, cause blockages at GG (and AG) and G, respectively, on the template strand,[42,44] whereas cyano-morpholinoadriamycin results in blockages at GG sites on either the template or nontemplate strands.[43] The monoadduct of psoralen is also known to block *E. coli* RNA polymerase when on the template strand, but not when on the nontemplate strand.[57,58]

It should be noted that the characteristics that are necessary to cause transcriptional blockage have yet to be defined at a molecular level. Major unresolved questions include

1. What is the minimum stereochemical appendage necessary to block the processive movement of *E. coli* RNA polymerase? For example, methylation of G-N7 does not cause blockages,[36]

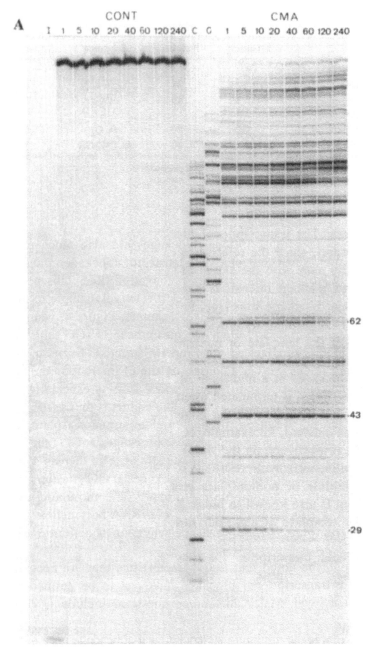

Figure 4.

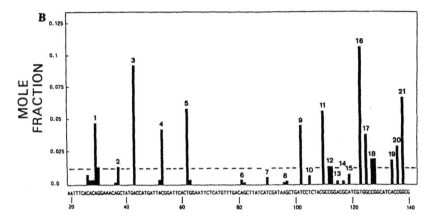

Figure 4. (Continued) Transcriptional blockages induced by cyano-morpholinoadriamycin.[49] (A) The initiated transcription complex was reacted with 1 μM cyanomorpholinoadriamycin (CMA) for 1 h, 37 °C, in transcription buffer, pH 8.0, and then elongated for 1–240 min prior to separation of transcripts by sequencing gel electrophoresis. The lane representing the initiated transcript is shown as I, and the 3′-methoxy-CTP and 3′-methoxy-GTP sequencing lanes are denoted as C and G. Control lanes of DNA not subjected to reaction with CMA, but elongated for 1–240 min, are denoted as CONT. (B) The sequence specificity of CMA was determined as the mole fraction of blocked transcripts after 1 min of elongation. The numbering is from G of the GpA dinucleotide used to initiate transcription. Of the 14 high-intensity blockage sites, 12 precede GG sequences on either the template or nontemplate strand.

    but the much larger nitrogen mustard adduct at the same site does result in blocked transcripts.

2. What is the effect of the site of attachment (i.e., minor groove vs. major groove, G-N2 vs. G-N7, etc.)?
3. What is the effect of attachment to template vs. nontemplate strands?
4. What is the effect of flexible rotation around the site of attachment?

What has clearly emerged is that, with the exception of methylation at G-N7 and A-N3, formation of permanent DNA adducts results in termination of transcription at, or immediately preceding, the adduct site.

C. Comparison to Other Methods

The drug sequence specificity detected from transcriptional blockages is generally the same as that detected by other methods [mainly DNase I and MPE-Fe(II) footprinting of reversibly binding drugs, and exonuclease digestion, primer-extension, and alkaline hydrolysis for adducts], as summarized in Tables 2 and 3. The major exception is the sequence specificity of the anthracyclines, adriamycin and daunomycin: the consensus sequence from the transcriptional assay was 5'-TCA, but was 5'-(A·T)(G·C)(G·C) from DNase I footprinting.[41,42] There are at least two possible sources for the apparent discrepancy. First, the transcriptional assay was at 10 °C (compared to 37 °C for DNase I footprinting), and, since there are only very small energy differences between several other preferred sites,[41,42] the temperature difference itself may be significant. Second, the effective drug loading is different in each case, and the population of occupied drug sites would therefore vary for each procedure.

The most precise definition of drug sites comes from alkaline hydrolysis of G-N7 adducts. For other procedures, the drug site is generally defined to within one base pair. Although the precision is comparable, the drug site is more readily obtained in the transcription assay than in classical footprinting, where a range of drug concentrations is usually required in order to yield the same level of definition of the footprint.

IV. RELATIVE OCCUPANCY

Each molecule of transcript contains the same amount of ^{32}P incorporated into the nascent RNA 10-mer (see Figures 1 and 2). The intensity of each band on the autoradiogram is therefore a measure of the number of molecules of RNA of that length. If the absorbance of individual bands is normalized with respect to the total intensity in each lane, that yields the mole fraction

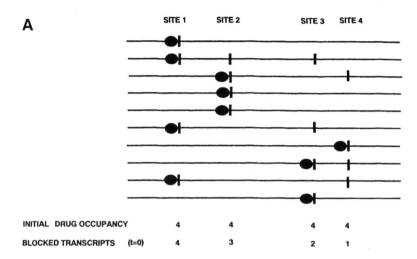

Figure 5. Diagrammatic representation of the relationship between actual drug occupancy and observed transcriptional blockages. The DNA molecules have been used to represent a population of initiated transcription complexes. An equal drug occupancy has been assumed at each of the four drug binding sites. Transcriptional blockages are implied when the processive movement of RNA polymerase is halted by a drug-occupied site (filled rectangle). These blockages are shown immediately after the elongation process has commenced (panel A), after 20 s of elongation (panel B), and after 200 s of elongation (panel C). The number of molecules of RNA polymerase reaching the downstream drug sites at early elongation times decreases progressively the further downstream; therefore, the underestimation of drug occupancy is greater at these sites. Once the drug dissociates from a blockage site, the RNA polymerase can move processively to the next drug-occupied site. Although the drug can then reoccupy such a vacated site, for simplicity these sites have been depicted as remaining unoccupied (open rectangle) once bypassed by the polymerase. With increasing elongation time, more RNA polymerase reaches downstream sites, which is reflected by an increase of blocked transcripts at those sites. The net effect of these processes is that the first drug site is accurately reflected by the mole fraction of blocked transcript, but all subsequent sites are progressively underestimated and are subjected to time-dependent opposing trends—an increase of blockages accompanying read-through of RNA polymerase past upstream sites, and a decrease of blockages as the drug dissociates from downstream sites.

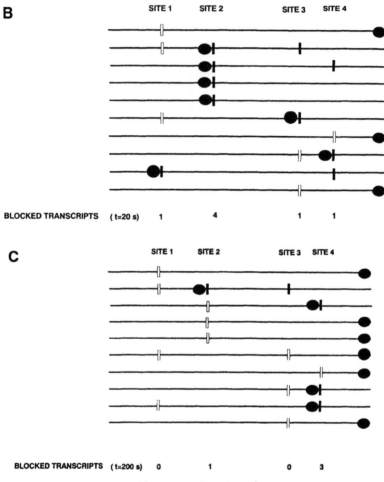

Figure 5. (Continued)

of RNA of that length. At low drug levels this will approximate the relative drug occupancy at each site. However, since RNA polymerase may not reach some downstream sites if there is high occupancy at all sites, the relative occupancy of these downstream sites can be substantially underestimated (this is represented diagrammatically in Figure 5). This underestimation does not of course apply to the first drug site but becomes increasingly serious further downstream. A simple correction for this statistical effect can be employed, based on the mole fraction of RNA of differing lengths (i.e., apparent drug occupancies at each site), A_i, but does not allow for drug dissociation rates.[21,28] The corrected relative concentration of RNA of one length ($A_{i,corr}$) is defined by the following equation:

$$A_{i,corr} = \frac{A_i}{(1 - \sum_{1}^{i-1} A_i)}$$

This expression includes a term for the probability that the ith site lies downstream of at least one of the $i-1$ upstream sites, which will block the movement of RNA polymerase.

By use of a simulation analysis (see Section VI), actual occupancy of all sites, as well as the kinetics of dissociation at each site,[22] can be calculated. A much simpler procedure is to use very low drug levels, such that, theoretically, only one drug site is occupied on each DNA molecule probed by the RNA polymerase. In practice, this would severely decrease the sensitivity of the method. However, to a good approximation, the level of blocked transcripts corresponds well to relative drug occupancy if at least 90% of full-length transcripts are observed.

V. DRUG DISSOCIATION KINETICS

It has been suggested that the effectiveness of drugs that act at the DNA level is related to the time that the drug resides on the DNA and, hence, upon the drug–DNA dissociation rate.[6,59,60] Simple physicochemical measurements of k_{off} (the rate constant for the dissociation of drug from DNA) may not reflect the situation in vivo, where DNA polymerase or RNA polymerase could significantly affect k_{off}.

The k_{off} of any DNA sequence specific drug can be readily determined under conditions of active transcription by measurement of the amount of blocked transcript at a drug site as a function of elongation time. The analysis of such data is simple only for the first drug site encountered by the RNA polymerase.

The gradual read-through of RNA polymerase past the first nogalamycin binding site (23-mer) is evident in Figure 3. The data for the decay of blocked transcripts at the first drug site has been well described by a first-order dissociation process for all drugs studied (see, for example, Figure 6). By this means, the time constant that was measured for dissociation of actinomycin D from the first 5′-GC site encountered by RNA polymerase was 2900 s.[20] The same site was synthesized as a 24-mer, with the GC site located in the middle of this sequence, and the actinomycin time constant that was determined, using SDS sequestration, was 140 s.[61] An explanation for the longer

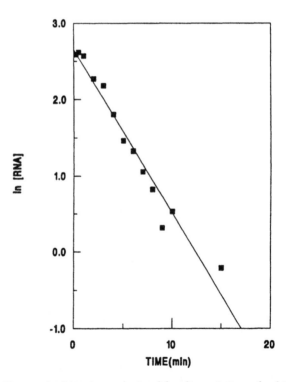

Figure 6. First-order kinetic analysis of the dissociation of echinomycin from the first major drug site (52-mer and 53-mer in Figure 7),[22] where there is a decreasing amount of blocked transcripts with increasing elongation time. The drug dissociation rate constant is 0.21 min^{-1}.

lived species under transcription conditions is that the polymerase has a "cage effect," trapping the drug in a restricted volume, thereby resulting in an effective higher free drug concentration and, hence, a longer drug residence time on the DNA.[61] There has not yet been any confirmation of this model.

VI. SIMULATION ANALYSIS OF MULTIPLE DRUG SITES

When a reversibly binding drug interacts at multiple sites with a DNA fragment, quantitative analysis of the blocked transcripts at each site yields true values of drug dissociation rate and occupancy only for the first drug site. This is because the observed amount of blocked transcript at downstream sites at early elongation times reflects that the polymerase is blocked at upstream sites, and therefore, the degree of blockage of downstream sites is underestimated. At later times, two components contribute to the amount of blocked transcript at a site: the rate of dissociation of drug from that site, which results in a loss of blocked transcript, and the extent of read-through from upstream sites, which results in an increase of polymerase reaching downstream sites. These effects are summarized in Figure 5.

In order to analyze rigorously the actual drug occupancy and dissociation rate at each site for a drug that binds reversibly to multiple sites, a simulation analysis is necessary. Two approaches have been employed, both yielding the same quantitative information, but with vastly differing degrees of ease. Both procedures have been applied to the analysis of 16 echinomycin binding sites for which the extent of movement of the RNA polymerase was measured at intervals over a 4h time span (Figure 7A).[22]

A. Monte Carlo Simulation

The Monte Carlo simulation[22] utilized a Fortran program to describe the rate of movement of RNA polymerase along the DNA. In essence, the DNA was represented as an array of 100,000 molecules, with one element for each drug site. This

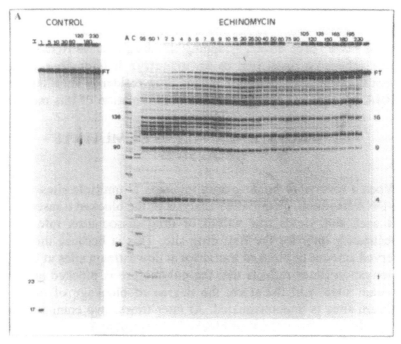

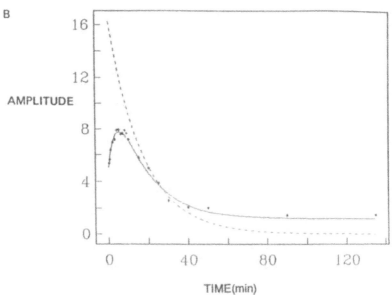

element was set to a nonzero value when the site was occupied or cleared when drug was absent. The rate of movement of RNA polymerase along each molecule was monitored as a function of time, with the probability that the polymerase would move from one occupied site to the next being proportional to the drug dissociation rate constant. The probability of termination at each site was calculated from a competing rate of termination of transcription at each site. The algorithm assumed that the RNA polymerase did not pause or terminate at empty drug sites, but moved rapidly from one drug site to the next compared to the rate of drug dissociation. All three values were varied until the decay of transcripts at the first drug site was fitted to the observed values. All other drug sites were then fitted in succession in the same manner. A complete quantitative picture of the echinomycin–DNA interaction resulted for each of the 16 sites from this analysis: 48 parameters were resolved, representing 16 relative drug occupancies, 16 drug dissociation rates, and 16 probabilities of induction of termination of transcription at that site.[22] The validity of this simulation is best illustrated by the good simulation of response for drug site 15, where a significant amount of read-through of RNA polymerase past earlier sites is evident (Figure 7B).

Figure 7. Kinetics of read-through past multiple echinomycin blockage sites.[22] (A) Echinomycin was reacted for 30 min, 37 °C with a 497-bp DNA fragment containing the initiated UV5 transcription complex (shown in lane I). Drug-induced blockages are shown after elongation for 0.25 to 230 min. The control lanes show the full-length transcript (FT, 380 residues) observed 1–230 min after elongation in the absence of echinomycin. Sequencing lanes A and C represent elongation of the initiated complex in the presence of 3′-methoxy ATP and 3′-methoxy CTP, respectively. Drug-induced blocked transcripts, which were subjected to simulation analysis, are numbered 1–16. (B) Monte Carlo simulation of echinomycin-induced blocked transcripts at echinomycin site 15 (Figure 7A), showing the experimental data (*), the simulated amount of observed blocked transcript (continuous line), and the simulated occupancy of drug at that site (dashed line).[22]

B. Compartmental Simulation

Although the Monte Carlo simulation was able to provide an excellent quantitative description of complex, multisite drug–DNA interactions, it is tedious and not suited to routine implementation and application in other laboratories. For these reasons an alternative procedure was developed, based on SAAM, the simulation analysis and modeling package.[23] In this simulation, RNA polymerase is either blocked at a drug site (with a probability defined by the drug occupancy at that site, and the drug is released at a rate defined by the drug dissociation rate constant at that site), or else it bypasses that site. This model is shown in Figure 8.

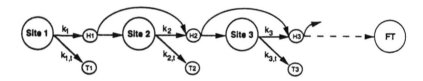

Figure 8. Simplified multicompartment model.[23] In this model, all RNA polymerase that escapes from drug site 2 (for example), as a result of drug dissociation from the DNA, is considered to be in a "holding" compartment (H2) from which there are only two possible progressions: it is either blocked at site 3 (the trapping efficiency being dependent solely on the drug occupancy at site 3), or else it bypasses site 3 at a virtually instantaneous rate. Possible termination of transcription at sites 1–16 is indicated as "termination" compartments T1–T16. The total amount of blocked transcript at any site (for example, site 2) is given by an equation that sums the amount of transcript at site 2 (comprising the amount of RNA that can be elongated from that site) and the amount that cannot elongate (T2) at any point of time in the elongation process. The rate of release of RNA polymerase past each site (the drug dissociation rate) is denoted as k_1, k_2, etc., and the competing rate, the rate of formation of terminated transcript, by $k_{1,t}$, $k_{2,t}$, etc. The probability of termination of transcript at any site is given by the fractional rate of movement into the termination compartment compared to the total rate of movement of RNA polymerase out of the dissociating drug compartment (i.e., $k_{1,t}/[k_{1,t} + k_1]$).

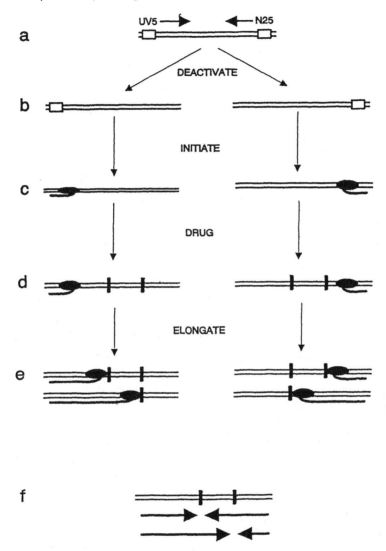

Figure 9. Schematic representation of bidirectional transcription foot-printing. The DNA fragment containing the counter-directed UV5 and N25 promoters is shown in (a). Selective deactivation of either one of the promoters yields (b) and addition of *E. coli* RNA polymerase and initiation nucleotides yields the appropriate initiated transcription complex (c). Reaction with drug yields (d) and subsequent elongation results in a range of drug-induced blocked transcripts (e). The two sets of blocked transcripts are summarized together in (f) to reveal bidirectional transcription footprints of drug sites.

The major advantages of this approach are that it is rapid (it can fit all 16 echinomycin sites simultaneously within 10 min on a 486DX desktop computer, compared to several months using the manual, site-by-site Monte Carlo approach)[23] and can be implemented rapidly by other laboratories. The parameters simulated by this procedure are essentially the same as those obtained by Monte Carlo simulation.

VII. BIDIRECTIONAL TRANSCRIPTION FOOTPRINTING

The in vitro transcription assay has been extended to yield a new technique, which we have termed "bidirectional transcription footprinting".[24] In the initial application of this method, two counter-directed promoters were separated by 100 bp. Each promoter was selectively deactivated by restriction digestion, and drug binding sites were detected by transcription from the remaining active promoter. The blocked transcripts therefore reveal both ends of the drug site and define a drug footprint. This method has been summarized diagrammatically in Figure 9. Application of this method to eight DNA-binding drugs has shown that the transcriptional footprint is revealed to be ±1 bp from both ends of the drug site (Table 4). The sensitivity and selectivity of this approach is such that two neighboring echinomycin CpG sites, separated by only 1 bp, have been readily resolved.

Figure 10. Bidirectional transcription footprinting of adriamycin-induced adducts (A) and adducts formed by cyanomorpholinoadriamycin (B). One promoter of the 315-bp DNA fragment was selectively deactivated by digestion with the appropriate restriction endonuclease. The initiated transcription complex was then formed from the remaining promoter and reacted with either adriamycin [10 μM, 24 h, in the presence of 75 μM Fe(III); panel A][30] or cyanomorpholinoadriamycin (2 μM, 1 h; panel B).[49] Elongation of the initiated complex was then carried out for 5 min, and the mole fraction of blocked transcripts was then quantitated for each promoter system and shown as a percentage of total transcripts at each blockage site.

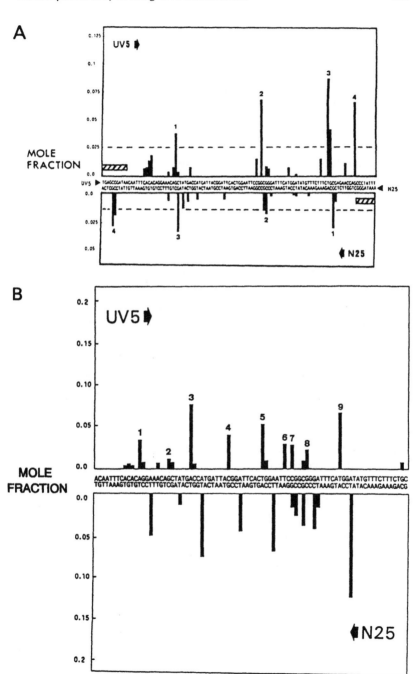

Table 4. Drug–DNA Sequence Specificity from Bidirectional
Transcription Footprinting

Drug	Sequence	Reference
Acridine–distamycin	$(A/T)_3$	24
Actinomycin D	GC	24
Adriamycin (adduct)	GC	30
Cisplatin	GG, GC	48
Cyanomorpholinoadriamycin	GG	49
Echinomycin	CG	24
Nogalamycin	CA	24
Mithramycin	GC	24

In contrast, it is difficult to obtain this information from DNase I footprinting on the same DNA fragment.[24]

A recent application of bidirectional transcription footprinting has been the analysis of DNA adducts induced by long-term exposure to the anticancer agent adriamycin to DNA. Four high-occupancy blockages were observed from each promoter (Figure 10A), and all were at G of GpC sequences of the nontemplate strand.[30] Three of the sites were detected by RNA polymerase from both promoters and revealed exceedingly clear footprints, which defined the adducts as being localized to the 2 bp element GpC. A highly cytotoxic derivative of adriamycin, cyanomorpholinoadriamycin (which readily forms adducts and interstrand cross-links with DNA), was also recently examined by bidirectional transcription footprinting.[49] Every GG sequence was revealed as a site of high occupancy (Figure 10B). It is therefore highly likely that this chemically reactive drug forms intrastrand cross-links at these GG sequences.

VIII. OTHER TRANSCRIPTION SYSTEMS

Almost all of the in vitro transcription studies of drug–DNA interactions have been performed by use of promoters for *E. coli* RNA polymerase (*lac* UV5, Tet^R, λP_L, and N25). This polymerase is large (molecular weight of approximately 450 kD) and complex (consisting of six subunits). Other smaller and less complex RNA polymerases have been tested to determine how they respond to the presence of drugs on the template DNA. The even more complex eukaryotic promoter systems have also been tested to determine if they could be utilized in the same way, given the additional complexity of a range of transcription factors required for promoter activity.

A. Bacteriophage RNA Polymerases

Three bacteriophage RNA polymerases have been utilized in the transcription assay (SP6, T3, and T7; see Table 1), and the sequence specificity and overall response has been studied with a total of seven different DNA-binding drugs.[21,28] Although the sequence specificity of drug sites was identical to that detected with the bacterial RNA polymerase, the bacteriophage RNA polymerases exhibited a dramatically enhanced tendency to induce termination of transcription at drug sites, and this termination appears to be dependent on the drug residence time.[21] Four of the drugs, with half-lives greater than 300 s, exhibited complete termination of transcription at every drug site, whereas for the other three drugs, with half-lives less than 300 s, no termination of transcription was observed.[21] These observations therefore raise the possibility that the probability of termination at a drug site may also prove to be dependent on drug residence time for the more complex RNA polymerases (bacterial and eukaryotic).

B. Eukaryotic Transcription Systems

Several laboratories have applied in vitro transcription systems to the study of drug–DNA interactions in recent years. Mithramycin has been shown to induce a kinetic blockage to RNA polymerase II,[62] whereas netropsin results in termination of transcription by RNA polymerase II.[63] A significant recent finding has been that elongation factor SII overcomes the transcriptional blockage of RNA polymerase II by distamycin.[64]

Overall, the in vitro eukaryotic transcription systems involving RNA polymerase II yield sequence specific blockages similar to those obtained with the bacterial promoter systems, a conclusion based upon the known preferred drug binding sites. However, the anticipated additional complexity of the eukaryotic system is graphically demonstrated by the effect of elongation factor SII in overcoming drug-induced blockages.[64] It is likely that an improved understanding of drug–DNA interactions under physiological conditions will emerge from studies based on these types of in vitro eukaryotic systems.

IX. CONCLUSIONS

The analysis of drug–DNA interactions by in vitro transcription analysis has provided new insight into these interactions. Phenomena such as termination at drug sites, drug residence time-dependent termination, delayed termination, and a possible cage effect of RNA polymerase have all been elucidated from the quantitation of these interactions at individual drug sites. In addition, bidirectional transcription footprinting has provided, with excellent resolution, a routine method for determining the precise location and physical size of small drugs and adducts on DNA. Further application of this method promises to provide additional new perspectives of drug–DNA interactions and as such can be expected to assist those involved in the design of more selective or effective DNA-acting agents, or those concerned with the mechanism of action of DNA-binding drugs. It is hoped that this technique can now be extended to more complex eukaryotic systems in vitro and, ultimately, to mammalian cells in vivo.

ACKNOWLEDGMENTS

The ideas, support, and encouragement of Professor Donald M. Crothers are gratefully acknowledged, as is the excellent developmental work of Dr. Robin White and Dr. Carleen Cullinane. Review and comments on this chapter by Dr. Cullinane are also much appreciated. The support of the Australian Research Council and the Anticancer Council of Victoria is also gratefully acknowledged.

REFERENCES

1. Schacter, L. P.; Anderson, C.; Canetta, R. M.; Kelley, S.; Nicaise, C.; Onetto, N.; Rozencweig, M.; Smaldone, L.; Winograd, B. *Seminars in Oncology* **1992**, *19*, 613–621.
2. Loxman, N. R.; Narayanan, V. L. *Chemical Structures of Interest to the Division of Cancer Treatment;* Bethesda: Drug Synthesis and Chemistry Branch, Developmental Therapeutics Program, National Cancer Institute, 1988; Vol. VI.
3. Chabner, B. A. In *Cancer: Principles and Practice of Oncology,* 4th ed.; DeVita, V. T., Hellman, S., Rosenberg, S.A., Eds.; Lippincott: Philadelphia, 1993; Chap. 18.
4. Dabrowiak, J. C. *Life Sciences* **1983**, *32*, 2915–2931.
5. Nielsen, P. E. *J. Molec. Recognition* **1990**, *3*, 1–25.
6. Leupin, W: In *Molecular Basis of Specificity in Nucleic Acid–Drug Interactions;* Pullman, B., Jortner, J., Eds.; Kluwer Academic: Dortrecht, 1990; pp 579–603.
7. Fox, K. R. In *Advances in DNA Sequence Specific Agents*; Hurley, L. H., Ed.; JAI, Greenwich, CT, 1992; Vol. 1, Chap. 5.
8. van Dyke, M. W.; Dervan, P. B. *Nucleic Acids Res.* **1983**, *11*, 5555–5567.
9. Fox, K. R.; Waring, M. J. *Nucleic Acids Res.* **1986**, *14*, 2001–2014.
10. Tullius, T. D. *Trends Biochem. Sci.* **1987**, *12*, 297–300.
11. Dabrowiak, J. C.; Goodisman, J. In *Chemistry and Physics of DNA–Ligand Interactions*; Kallinbach, N. R., Ed.; Academic: New York, 1989; pp 143–174.
12. Goodisman, J.; Dabrowiak, J. C. In *Advances in DNA Sequence Specific Agents;* Hurley, L. H., Ed.; JAI Press: Greenwich, CT, 1992; Vol. 1, Chap. 2.
13. Robbie, M.; Wilkins, R. J. *Chem.-Biol. Interact.* **1984**, *49*, 189–209.
14. Murray, V.; Motyka, H.; England, P. R.; Wickham, G.; Lee, H. O.; Denny, W. A.; McFadyen, W. D. *J. Biol. Chem.* **1992**, *267*, 18805–18809.
15. Murray, V.; Motyka, H.; England, P. R.; Wickham, G.; Lee, H. H.; Denny, W. A.; McFadyen, W. D. *Biochemistry* **1992**, *31*, 11812–11817.
16. Aivashahvilli, V. A.; Beabeahashvilli, R. Sh. *FEBS Lett.* **1983**, *160*, 124–128.
17. Phillips, D. R.; Crothers, D. M. *Biochemistry* **1986**, *25*, 7355–7362.
18. Straney, D. C.; Crothers, D. M. *Cell* **1985**, *43*, 449–459.
19. von Hippel, P. H.; Bear, D. G.; Morgan, W. D.; McSwiggen, J. A. *Ann. Rev. Biochem.* **1984**, *53*, 389–445.
20. White, R. J.; Phillips, D. R. *Biochemistry* **1988**, *27*, 9122–9132.
21. White, R. J.; Phillips, D. R. *Biochemistry* **1988**, *28*, 4277–4283.
22. Phillips, D. R.; White, R. J.; Dean, D.; Crothers, D. M. *Biochemistry* **1990**, *29*, 4812–4819.
23. Phillips, D. R.; Moate, P. J.; Boston, R. C. *Anti-Cancer Drug Design* **1994**, *9*, 209–219.
24. White, R. J.; Phillips, D. R. *Biochemistry* **1989**, *28*, 6259–6269.
25. Phillips, D. R.; Crothers, D. M. In *Methods in Molecular Biology*; Humana Press: NJ, 1994; Vol. 37, Chap. 7.
26. Phillips, D. R.; White, R. J.; Trist, H.; Cullinane, C.; Dean, D.; Crothers, D. M. *Anti-Cancer Drug Design* **1990**, *5*, 21–29.

27. Phillips, D. R.; Cullinane, C.; Trist, H.; White, R. J. In *Molecular Basis of Specificity in Drug–DNA Interactions;* Pullman, B., Jortner, J., Eds.; Kluwer Academic: Dortrecht, 1990; pp 137–155.

28. Trist, H.; Phillips, D. R. *Nucleic Acids Res.* **1989**, *17*, 3673–3688.

29. Carpousis, A. J.; Stefano, J. E.; Gralla, J. D. *J. Mol. Biol.* **1982**, *157*, 619–633.

30. Cullinane, C.; Phillips, D. R. *Biochemistry* **1990**, *29*, 5638–5646.

31. Stefano, J. E.; Gralla, J. D. *Biochemistry* **1979**, *18*, 1063–1067.

32. Dabrowiak, J. C.; Skorobogaty, A.; Vary, C. P. H.; Vournakis, J. N. *Nucleic Acids Res.* **1986**, *14*, 489–499.

33. Zhou, W.; Doetsch, P. W. *Proc. Natl. Acad. Sci. U.S.A.* **1993**, *90*, 6601–6605.

34. Masta, A.; Gray, P. J.; Phillips, D. R., *Nucleic Acids Res.* **1994**, *22*, 3880–3886.

35. Melton, D. A.; Krieg, P. A.; Rebagliati, M. R.; Maniatis, T.; Zinn, K.; Green, M. R. *Nucleic Acids Res.* **1984**, *12*, 7035–7053.

36. Cullinane, C. Ph.D. Thesis, La Trobe University, 1994, p. 60.

37. Skorobogaty, A.; White, R. J.; Phillips, D. R.; Reiss, J. A. *FEBS Letters* **1988**, *227*, 103–106.

38. Skorobogaty, A.; White, R. J.; Phillips, D. R.; Reiss, J. A. *Drug Design Deliv.* **1988**, *3*, 125–152.

39. Panousis, C.; Phillips, D. R. *Nucleic Acids Res.* **1994**, *22*, 1342–1345.

40. Chen, F. *Biochemistry* **1988**, *27*, 1843–1848.

41. Chaires, J. B.; Fox, K. R.; Herrera, J. E.; Britt, M.; Waring, M. J. *Biochemistry* **1987**, *26*, 8227–8236.

42. Chaires, J. B. In *Advances in DNA Sequence Specific Agents*; Hurley, L. H., Ed.; JAI Press: Greenwich, CT, 1992; Vol. 1, Chap. 1.

43. Low, M. L.; Drew, H. R.; Waring, M. J. *Nucleic Acids Res.* **1984**, *12*, 4865–4879.

44. van Dyke, M. W.; Dervan, P. B. *Science* **1984**, *225*, 1122–1127.

45. Fox, K. R.; Waring, M. J. *Biochemistry* **1986**, *25*, 4349–4357.

46. Lybrand, T. B.; Brown, S. C.; Creighton, S.; Shafer, R. H.; Kellman, P. A. *J. Mol. Biol.* **1986**, *191*, 495–507.

47. Phillips, D. R.; White, R. J.; Cullinane, C. *FEBS Lett.* **1989**, *246*, 233–240.

48. Cullinane, C.; Wickham, G.; McFadyen, W. D.; Denny, W. A.; Palmer, B. D.; Phillips, D. R. *Nucleic Acids Res.* **1993**, *21*, 393–400.

49. Cullinane, C.; Phillips, D. R. *Biochemistry* **1992**, *31*, 9513–9519.

50. Gray, P. J.; Cullinane, C.; Phillips, D. R. *Biochemistry* **1991**, *30*, 8036–8040.

51. Lemaire, M. A.; Schwartz, A.; Rahmouni, A. R.; Leng, M. *Proc. Natl. Acad. Sci. U.S.A.* **1991**, *88*, 1982–1985.

52. Mattes, W. B.; Hartley, J. A.; Kohn, K. W. *Nucleic Acids Res.* **1986**, *14*, 2971–2987.

53. Kohn, K. W.; Hartley, J. A.; Mattes, W. B. *Nucleic Acids Res.* **1987**, *15*, 10531–10549.

54. Borowy-Borowski, H.; Lipman, R.; Tomasz, M. *Biochemistry* **1990**, *29*, 2999–3006.

55. Tomasz, M.; Borowy-Borowski, H.; McGuinness, B. F. In *Molecular Basis of Specificity in Nucleic Acid–Drug Interactions;* Pullman, B., Jortner, J., Eds.; Kluwer Academic: Dortrecht, 1990; pp 551–564.

56. Bizamek, R.; McGuinness, B. F.; Nakaniski, K.; Tomasz, M. *Biochemistry* **1992**, *31*, 3084–3091.

57. Shi, Y.-B.; Gamper, H.; Hearst, J. E. *Nucleic Acids Res.* **1987**, *15*, 6843–6854.

58. Selby, C.; Sancar, A. *Science* **1993**, *260*, 53–58.

59. Feigon, J.; Denny, W. A.; Leupin, W.; Kearns, D. R. *J. Med. Chem.* **1984**, *27*, 450–465.

60. Wakelin, L. P. G.; Atwell, G. J.; Rewcastle, G. W.; Denny, W. A. *J. Med. Chem.* **1987**, *30*, 855–861.

61. White, R. J.; Phillips, D. R. *Biochem. Pharmacol.* **1989**, *38*, 331–334.

62. Hardenbol, P.; Van Dyke, M. W. *Biochem. Biophys. Res. Commun.* **1992**, *185*, 553–558.

63. Ueno, A.; Baek, K.; Jeon, C.; Agarwal, K. *Proc. Natl. Acad. Sci. U.S.A.* **1992**, *89*, 3676–3680.

64. Mote, J.; Ghanouni, P.; Reines, D. *J. Mol. Biol.* **1994**, *236*, 725–737.

PART II

SEQUENCE SPECIFICITY OF DNA INTERACTIVE DRUGS

MOLECULAR RECOGNITION OF DNA BY DAUNORUBICIN

Jonathan B. Chaires

Advances in DNA Sequence Specific Agents
Volume 2, pages 141–167.
Copyright © 1996 by JAI Press Inc.
All rights of reproduction in any form reserved.
ISBN: 1-55938-166-3

141

I. INTRODUCTION

The anthracycline antibiotics, of which daunorubicin and doxorubicin (Figure 1) are the parent compounds, are the class of antitumor drugs with the widest spectrum of activity in human cancers.[1] As recently as 1988, doxorubicin was the leading anticancer drug in terms of sales in the United States.[2] Clinical trials of daunorubicin began in 1964, after its discovery and isolation as a natural product from a species of *Streptomyces*. In spite of exhaustive attempts to synthesize chemically derivatives of daunorubicin or doxorubicin with greater clinical potency or lesser toxicity, no compounds have yet been found that offer any substantial improvement over the original natural product.[1] Because the anthracyclines have been, and continue to be, among the most potent weapons in the chemical arsenal used in cancer chemotherapy, their chemistry, biochemistry, and pharmacology have been extensively studied over the last three decades. Several books[3–6] and numerous reviews[7–21] have described and summarized the results of such studies through the mid 1980s. The last decade has seen significant advances in the understanding of both the structures of crystalline complexes

Anatomy of a Daunorubicin Molecule

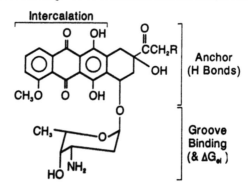

Figure 1. Structure of daunorubicin. For daunorubicin, R = H, and for doxorubicin, R = OH. The three functional domains of the molecule are indicated.

of anthracycline antibiotics with DNA and the properties of anthracycline–DNA complexes in solution. Particularly important advances have led to clear delineation of the sequence specificity and structural consequences resulting from the binding of these drugs to DNA. Indeed, X-ray crystallographic studies, molecular modeling studies, and physicochemical studies have converged to make the daunorubicin–DNA interaction perhaps the best understood of the intercalation reactions and, therefore, an important fundamental model on which to base attempts aimed toward the rational design of DNA-targeted drugs.

The cytotoxic effects of daunorubicin and doxorubicin are pleiotropic and may involve interactions with DNA or with membranes,[22] or the production of free radicals.[23] There exists compelling evidence to indicate that cellular DNA is a primary target for the anthracycline antibiotics. In a study of 26 anthracycline derivatives, a direct correlation was found between their DNA-binding affinity and bioactivity.[24] Quantitative microspectrofluorometry was used to show that doxorubicin is rapidly accumulated into the nuclei of living cells, where it is nearly completely bound to the DNA.[25] Fluorescence resonance energy transfer methods were used in flow cytometry experiments to show that daunorubicin is intercalated into DNA within living cells.[26] The surface-enhanced Raman spectrum of doxorubicin within the nuclei of living erythroleukemic cancer cells was found to resemble the spectrum obtained in vitro for the drug complexed with DNA.[27] In a provocative study, doxorubicin was found to displace selectively a distinct set of nuclear proteins.[28] All of these studies indicate that, in vivo, anthracycline antibiotics are strongly associated with cellular DNA. Over the last decade, topoisomerase II has been identified as a key target for a variety of anticancer drugs, including the anthracyclines.[29] Although the exact molecular mechanism by which the anthracyclines poison this key enzyme remains poorly defined, recent studies have shown that DNA binding is necessary for the inhibition.[30,31]

The primary purpose of this chapter is to review the progress made in understanding the molecular recognition of DNA by daunorubicin, the prototype anthracycline antibiotic. It is now clear that daunorubicin shows a strong structural specificity for right-handed, B-form DNA. In addition, daunorubicin shows a distinct, and somewhat unusual, sequence preference in its binding to DNA. Although the sequence specificity of its DNA is by no means absolute, daunorubicin preferentially recognizes the triplet motifs 5'(A/T)GC and 5'(A/T)CG, where the notation (A/T) indicates that either A or T may occupy the sequence position. Given the rich background of structural data for the anthracycline antibiotics, it is possible to rationalize these structural and sequence preferences and to specify in detail the probable molecular determinants of the interaction of daunorubicin with particular DNA sites. Daunorubicin represents an important, simple system from which to learn design principles for the molecular recognition of DNA by small molecules.

II. STRUCTURAL STUDIES OF THE DAUNORUBICIN–DNA COMPLEX

Table 1 summarizes high resolution X-ray crystallographic studies of anthracycline antibiotic–DNA complexes.[32–46] Inspection of the table shows that a wealth of structural data is available for the anthracyclines. Structures of complexes of daunorubicin bound to several different sequences have been obtained. Conversely, the structures of several different anthracycline antibiotics bound to the same DNA sequence have been solved. For no other intercalator does such a variety of structural information exist.

These structural studies reveal several common features of anthracycline binding to DNA. First, in all studies, the intercalation of the anthraquinone ring system is evident. The long axis of the drug chromophore is nearly perpendicular to the long axis of the DNA base pairs, a feature that contrasts with the structures of DNA complexes of simple intercalators like ethidium and proflavin, in which the long axes of the ligand and DNA base pairs are nearly parallel. Second, again in contrast to the simple intercalators, several hydrogen-bonding interactions appear to stabilize anthracycline–DNA complexes. In all of the observed structures listed in Table 1, two types of hydrogen-

Table 1. Summary of Published X-Ray Crystallographic Studies of Anthracycline Antibiotic–Deoxyoligonucleotide Complexes

Sequence	Drug:DNA	Resolution	R Factor	Ref.
I. *Daunorubicin*				
a. 5′ CGTACG	2:1	1.54	0.2	32
b. 5′ CGTACG	2:1	1.2	0.175	33
c. 5′ CGTACG	2:1	1.2	0.171	34
d. 5′ CGATCG	2:1	1.5	0.25	35
e. 5′ CGATCG	2:1	1.5	0.175	36
f. 5′ TGTACA	2:1	1.6	0.25	37
g. 5′ TGATCA	2:1	1.7	0.26	37
h. 5′ CGCGCG	1:1[a]	1.5	0.185	38
i. 5′ CGTDCG	1:1[a]	1.5	0.176	38
j. 5′ CGCGCG	2:1	1.9	0.21	39
k. 5′ CGATCG	2:1	1.2	0.271	40
II. *Doxorubicin*				
a. 5′ CGATCG	2:1	1.7	0.177	36
b. 5′ TGGCCA	2:1	1.7	0.22	39
c. 5′ CGATCG	2:1	1.2	0.265	40
III. *4′-Epiadriamycin*				
a. 5′ CGATCG	2:1	1.5	0.196	41
b. 5′ TGTACA	2:1	1.4	0.17	42
c. 5′ TGATCA	2:1	1.7	0.202	43
IV. *Idarubicin*				
a. 5′ CGATCG	2:1	1.7	0.188	44
b. 5′ TGATCA	2:1	1.6	0.22	45
V. *4-O-Demethyl-11-deoxydoxorubicin*				
a. 5′ CGATCG	2:1	1.8	0.179	44
VI. *Morpholinodoxorubicin*				
a. 5′ CGTACG	2:1	1.5	0.192	46

Note: [a]Covalently cross-linked drug–DNA complex.

bonding interactions have been seen. In the first of these, two direct hydrogen bonds between the 9-hydroxy group on the drug and the N-2 and N-3 residues on the central guanine are observed. In the second case, a *solvent-mediated* hydrogen bond between ligand groups O-4/O-5 and purine N-7 (of the upper base pair in the intercalation site) is seen.

In all structures reported so far, the daunosamine extends from the intercalation site and fits snugly in the minor groove. However, the exact position of the sugar is variable, and whether or not the amine moiety participates in hydrogen bond interactions appears to depend on the DNA sequence. If the central dinucleotide sequence is 5'TA (e.g., entries Ia–c, f, Table 1), the amine moiety of the daunosamine *does not* form any direct hydrogen bonds within the minor groove. However, if the central sequence is reversed to 5'AT (e.g., entries Id–g, k, Table 1), the amine drops to within direct hydrogen-bonding distance to atoms on the floor of the minor groove. Although in all cases the daunosamine is tightly bound in the minor groove, its exact position appears to be sequence-dependent, as does its participation in hydrogen-bonding interactions.

The structural studies, in their most general interpretation, indicate that the anthracyclines are *polyfunctional* DNA binding ligands, as shown in Figure 1. These ligands are tripartite. The anthraquinone ring system *intercalates* and stabilizes the complex by stacking interactions within the intercalation site. The A ring acts as an *anchor* and stabilizes the complex by both direct and indirect (solvent-mediated) hydrogen-bonding interactions. The daunosamine acts as a tiny *groove binder* and stabilizes the complex in at least two different ways. In addition to van der Waals interactions within the minor groove, the daunosamine may provide additional sequence-dependent hydrogen-bonding interactions between its amine group and DNA. Further, because the daunosamine carries the single positive charge on daunorubicin, it makes an electrostatic contribution to the overall binding free energy.

Several NMR studies have been published that indicate that the structural features observed in crystals appear to persist in solution.[47–55] Early studies of daunorubicin complexed to deoxypolynucleotides revealed an intercalative binding mode and the perpendicular orientation of the anthraquinone ring system relative to the long axis of the DNA base pairs. None of these studies, except one,[55] have attempted high-resolution structural determination of the drug–DNA complexes, but they do provide structural information that is consistent with the results of X-ray crystallographic studies.

A limited number of vibrational spectroscopic studies (Raman and infrared) have appeared that characterized the complexes of either daunorubicin or doxorubicin with DNA.[56–58] These studies provide indications of additional molecular interactions not readily evident in X-ray crystallographic or NMR studies. First, resonance Raman studies indicate possible hydrogen-bonding interactions of the phenolic groups on the anthraquinone ring with DNA bases within the intercalation site.[56] Second, an IR spectroscopic study shows that the doxorubicin C-14 hydroxyl may hydrogen bond with DNA phosphates.[58] Such an interaction would be generally consistent with the slightly higher DNA-binding affinity of doxorubicin relative to daunorubicin.

In summary, structural studies show that anthracycline antibiotics bind to DNA by a unique mixed mode, with both an intercalating portion and a groove-binding portion. Apart from the stacking interactions within the intercalation site, and the van der Waals interactions within the minor groove, several hydrogen-bonding interactions stabilize the drug–DNA complexes. The presence of such hydrogen bonds distinguishes daunorubicin from the simple intercalators like ethidium and proflavin, which cannot form such bonds when bound to DNA. As will be seen, these hydrogen bond interactions contribute to the sequence specificity of DNA binding of the anthracyclines.

III. COMPUTATIONAL STUDIES OF ANTHRACYCLINE–DNA COMPLEXES

Key insights into the anthracycline–DNA interaction emerged from computational studies performed in the laboratory of Professor Bernard Pullman.[8,56–63] These computational studies are of particular interest and significance in that they predicted *in advance of definitive experimental studies* the sequence selectivity of daunorubicin binding to DNA. (See Reference 63 for a particularly lively and entertaining discussion of this point.)

Computational studies showed first of all that daunorubicin recognizes a *triplet* sequence. This finding is consistent with structural studies, which show that the drug physically covers 3 base pairs. The energetically most favorable triplet was found to be 5'TCG, followed (in order) by 5'ACG > 5'ATA > 5'ACI 5'GCG > 5'GTA. The basis of the sequence preference is due to hydrogen-bonding interactions and a surprising contribution of the daunosamine. The C-9 hydroxyl of daunorubicin forms a particularly strong double hydrogen bond with the N-2 and N-3 of the guanine at the center of the preferred triplet binding site. (Recall that this interaction is observed in all of the structures listed in Table 1.) If adenine were substituted for guanine, only a single hydrogen bond can form. Thus, there is an energetic preference for guanine at the center of the binding site. The preference for an A or T at the 5' terminus is more subtle. The daunosamine lies snugly in the minor groove, extending away from the intercalation site and physically covering a third base pair. An AT base pair is preferred at this position because it allows a better stereochemical fit of the daunosamine within the minor groove. If a GC base pair were to occupy the 5'-terminal position, the N-2 of guanine would protrude into the minor groove and sterically hinder the optimal fit of the daunosamine.

A few computational studies followed those of the Pullman laboratory.[64] Free energy perturbation methods were used to arrive at the same concluded sequence preference.[64]

IV. MACROSCOPIC BINDING STUDIES OF THE DAUNORUBICIN–DNA INTERACTION

Figure 2 shows a binding isotherm for the interaction of daunorubicin with calf thymus DNA. This figure is a composite and was constructed by a combination of data from four different laboratories that had studied the binding reaction under identical solution conditions[65–69] (S. Satyanarayana, J. B. Chaires, unpublished data). A variety of experimental approaches have been used to obtain these data, including fluorescence and absorption

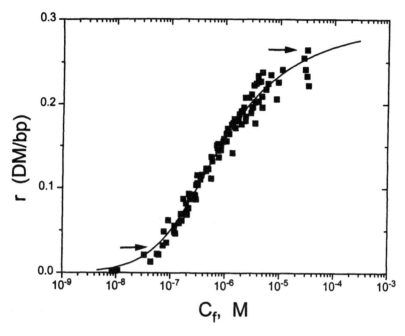

Figure 2. Composite binding isotherm for the interaction of daunorubicin with calf thymus DNA at 20 °C, pH 7.0, 0.2 M Na⁺. Data were taken from reports from four different laboratories.[65–69] Data were fit to the neighbor exclusion model,[70] yielding the best fit shown as the solid line.

spectroscopy, equilibrium dialysis, and phase partition methods. It is clear that the combined data shown in Figure 2 are in excellent agreement. Analysis of the data of Figure 2 by the simple neighbor exclusion model of McGhee and von Hippel[70] yields a binding constant (K) for the interaction of daunorubicin with an isolated DNA site of 6.6 (± 0.2) $\times$ 10⁵ M⁻¹ and an exclusion parameter (n) of 3.3 $\pm$ 0.1 bp. The latter value is consistent with structural studies (Table 1) that show that daunorubicin physically covers 3 bp when bound to DNA. It must be emphasized that the binding isotherm shown in Figure 1 is a *macroscopic* characterization of binding and the K value obtained by the analysis using the neighbor exclusion model is

Table 2. Thermodynamic Profile for the Interaction of Daunorubicin
with Calf Thymus DNA[a]

K = 6.9 (±0.2) × 10^5 M^{-1}
n = 3.3 (±0.1) bp
ΔG^0 = $-RT \ln K$ = −7.8 kcal mol^{-1}
ΔH^0 = −13.3 (±0.6) kcal mol^{-1} (van't Hoff)
−10.4 (±1.1) kcal mol^{-1} (calorimetry)
ΔS^0 = −8.9 cal mol^{-1} deg^{-1}
($\delta \ln K / \delta \ln$ [Na^+]) = −1.25 ± 0.2
ΔG_{obs} = $\Delta G_0 + \Delta G_{el}$ = −6.9 + (0.738) ln [Na^+]
K_{obs} = (37.0 × 10^5)(2$f^2 - f^3$) where f = fraction GC bp

Note: [a]The values reported refer to standard solution conditions of 0.2 M NaCl, pH 7.0., 20 °C.

obtained under the assumption of identical and noninteracting binding sites along the DNA lattice. Footprinting studies, to be summarized in Section VI, reveal that this assumption is probably invalid. The macroscopic K value is, nonetheless, still a useful quantitative measure of binding, if it is kept in mind that it represents a complicated average binding constant that masks a distribution of binding constants for the interaction of daunorubicin at different sequences along the DNA lattice.

By a study of daunorubicin binding to DNA as a function of temperature, ionic strength, and base composition, it is possible to derive a complete *macroscopic* thermodynamic profile for the binding interaction. Table 2 shows such a profile. Several points emerge from Table 2. Daunorubicin binding to calf thymus DNA is energetically favorable, with ΔG^0 = −7.8 kcal mol^{-1}. The favorable binding free energy results largely from a favorable binding enthalpy (ΔH^0= −10.4 kcal mol^{-1}). The direct measurement, by calorimetry, of the binding enthalpy of daunorubicin for its interaction with a number of natural DNA samples and polynucleotides is an area of recent progress.[71,72] The negative entropy that accompanies daunorubicin binding must arise from the complicated interplay of many contributions, including changes in both DNA and antibiotic hydration, ion release, the loss of translational and rotational freedom by the antibiotic upon binding, and DNA conformational changes upon intercalation.

Table 2 also shows that daunorubicin binding to DNA is strongly dependent upon NaCl concentration, as reflected by the quantity ($\delta\ln K/\delta\ln$ [Na$^+$]). This salt dependence has been analyzed in detail.[73] Briefly, the observed salt dependence arises from the coupling of the binding of positively charged drug molecules with Na$^+$ binding to DNA. The overall contribution to the binding free energy arising from polyelectrolyte effects is comparatively minor, approximately −1.0 kcal in 0.2 M NaCl. The implication of this is that the main noncovalent interactions stabilizing the daunorubicin–DNA complex are hydrogen bonds and van der Waals interactions.

Finally, Table 2 shows that the magnitude of the daunorubicin binding constant for DNA is strongly dependent upon the base composition of the DNA to which it is binding. The functional dependency is complex and goes as the *square* of the fractional GC content. Such a dependency reflects site specific drug binding and may be explained by a model in which daunorubicin preferentially binds to sites containing adjacent GC base pairs.[74]

The kinetics of daunorubicin binding to DNA have been thoroughly studied.[75–80] These studies reveal that daunorubicin binding to DNA is both slow and complex. The lifetime of the daunorubicin–DNA complex is approximately one second, and at least three steps are required to describe the time course of its DNA binding reaction. These features distinguish daunorubicin from simple intercalators like ethidium and proflavin, whose binding is complete on the millisecond timescale and which seem to bind by a simple bimolecular reaction mechanism. The complicated binding kinetics exhibited by daunorubicin most probably arise from its preferential binding to certain DNA sequences.[10]

V. STRUCTURAL SPECIFICITY OF THE DAUNORUBICIN–DNA INTERACTION

Daunorubicin and doxorubicin both show a strong *structural specificity* in their interaction with DNA. Right-handed B-form

DNA is strongly preferred as a binding site over a variety of alternate DNA conformations that have been examined, including left-handed Z-DNA,[81-88] the unusual right-handed helical form adopted by poly(dA)·poly(dT),[89] and DNA that has been wrapped into nucleosomes.[90-93]

Both daunorubicin and doxorubicin were found to inhibit the rate of the NaCl-induced B–Z transition in poly(dGdC).[81,82] Daunorubicin, in addition, was found to increase the transition midpoint of the thermally driven B–Z transition in poly-(dGm^{5}dC).[87] Most interestingly, however, daunorubicin was shown to act as an allosteric effector of DNA conformation and was found to convert Z-DNA to an intercalated right-handed form under solution conditions that would otherwise strongly favor the Z form of the DNA. ^{31}P NMR,[83] circular dichroism,[83-85] sedimentation velocity,[84] and susceptibility to DNase I digestion[84] have all been used to demonstrate the Z–B conversion induced by daunorubicin. The coupling of daunorubicin binding to the Z–B conformational transition gives rise to positive co-operativity in binding isotherms under solution conditions that initially favor the Z form. An example is shown in Figure 3. Such binding data may be fit to an allosteric model that includes as adjustable parameters binding constants for the interaction of daunorubicin with both Z- and B-DNA and nucleation and propagation constants for the Z–B transition.[86] The allosteric model is analogous to the well-known Monod–Wyman–Changeaux model applied to cooperative protein binding interactions but differs in the mechanistic details of the conformational change in the macromolecule. The energetics of the interaction of daunorubicin with left-handed poly(dGdC) in 3.5 M NaCl are summarized in Figure 3. The free energy profile shows that, although the Z–B transition is energetically unfavorable under these conditions ($\Delta G^0 > 0$), drug binding is much more favorable to the B form compared to the Z form. As a result, daunorubicin binding facilitates the conversion of Z-DNA to the B form. The binding constant of daunorubicin to B-DNA is 45 times higher than the binding constant to the Z form under these ionic conditions, a difference that corresponds to over a 2 kcal mol^{-1} more favorable binding free energy. Daunorubicin

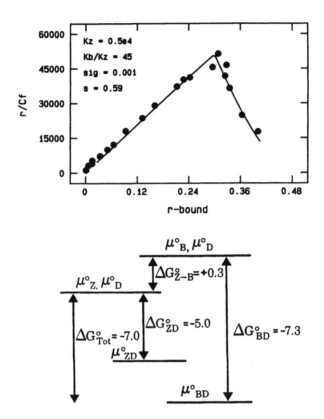

Figure 3. Allosteric conversion of Z-DNA to a right-handed intercalated form by daunorubicin. Data are taken from Reference 86. The top part shows the cooperative binding isotherms found for the interaction of daunorubicin with poly(dGdC) in 3.5 M NaCl, a salt concentration that initially favors the left-handed Z form of the polynucleotide. The solid line is the fit of the experimental data to an allosteric model in which drug binding is coupled to the Z–B conformational transition in the polynucleotide.[86] The lower part shows a free energy diagram for the allosteric system. The chemical potential of a base pair in either the Z or B conformation is denoted μ_Z^0 and μ_B^0, respectively. The chemical potential of free daunorubicin is μ_D^0, and its complexes with Z- or B-DNA are μ_{ZD}^0 and μ_{BD}^0, respectively. The higher affinity of daunorubicin for the B form of poly(dGdC) facilitates the Z–B conversion of the polymer and is the energetic driving force for the conversion of Z-DNA to the intercalated, right-handed form.

thus exhibits a striking specificity for right-handed DNA and effectively discriminates against the left-handed Z form.

A more subtle conformational discrimination is shown by daunorubicin in its interaction with poly(dA)·poly(dT).[89] Poly(dA)·poly(dT) adopts an unusual right-handed conformation distinct from the standard B form. A thermally driven transition can occur to drive the polymer into a form more like standard B-DNA, with concomitant changes in the UV absorption and circular dichroism spectra.[89] Daunorubicin binding to poly(dA)· poly(dT) exhibits positive cooperativity.[74] The allosteric model mentioned above may be used to show that daunorubicin binds more tightly to the more standard B form of the polynucleotide, with a binding constant four times higher than that characteristic of binding to the nonstandard right-handed form of the polymer. Binding is tightly coupled to the helix-to-helix conformational transition of poly(dA)·poly(dT), as demonstrated by circular di-chroism and the susceptibility to DNase I digestion.[89] Daunoru-bicin thus discriminates against the nonstandard right-handed form of poly(dA)·poly(dT) and can allosterically convert the polymer to a form with higher binding affinity. The allosteric conversion is entirely analogous to that observed in the case of Z-DNA, but the magnitude of the energetics involved differ considerably. The preference of daunorubicin for B- over Z-DNA is much more pronounced than its preference for one right-handed form of poly(dA)·poly(dT) over the other.

Daunorubicin binding to nucleosomal DNA was found to be strongly reduced relative to its affinity for free DNA.[90] Further studies showed that other anthracycline antibiotics, including doxorubicin, behaved in the same way.[91,93] The reduced binding affinity for nucleosomal DNA was proposed to result from its bent conformation.[90] The preference of daunorubicin for free DNA suggested that it might accumulate in genetically active regions of DNA within the nucleus, in which nucleosomal struc-ture is less prevalent.[90] The interaction of daunorubicin with nucleosomes is, however, complex.[90] Binding of the drug un-folds nucleosomes but does so without causing dissociation of the histones from the DNA. At binding ratios of greater than 0.15 mol drug per DNA base pair, daunorubicin promotes nucleosomal aggregation. In spite of these dramatic effects on nucleosome conformation, however, DNase I footprinting studies show only subtle differences between the interaction of daunorubicin with the free *tyr T* DNA fragment and the same fragment reconstituted into a nucleosome.[92] No shift in the sequence preference in the binding interaction is evident in these studies. A few sites do, however, show enhanced cleavage in the nucleosomal DNA upon daunorubicin binding, but not in the free DNA. The origin of this effect is not clear.

In all cases that have been examined in detail, conversion of DNA to a form different from the standard right-handed B conformation results in a dramatically reduced affinity for daunorubicin. Daunorubicin thus shows a strong structural specificity and will discriminate against non-B-form conformations.

VI. SEQUENCE SPECIFICITY OF THE DAUNORUBICIN–DNA INTERACTION

The advent of "footprinting" methodology[94–98] has made it possible to study the *microscopic* binding of antibiotics to specific DNA sites. Both DNase I footprinting[99–103] and a high-resolution transcription assay[104,105] have been used to identify preferred anthracycline antibiotic binding sites in DNA. The results from these methods reveal unambiguously that the anthracyclines do not bind randomly along the DNA lattice, but rather bind preferentially to particular DNA sequences. Aspects of quantitative footprinting titration studies of daunorubicin will be highlighted here.

Footprinting titration studies of the interaction of daunorubicin with a 165-bp restriction fragment containing the tyrosine tRNA operator region (the "*tyr T* fragment") revealed four responses, corresponding to four different classes of sites, upon addition of drug.[102] One class of sites showed no effect of the antibiotic on the DNase I cleavage rate. Such sites are interpreted as sites that do not bind the antibiotic. A second class of sites showed protection from DNase I cleavage upon addition of drug and represents sites near where the antibiotic binds. Another class of sites showed protection but only after a threshold drug concentration had been reached. These sites may arise from cooperative binding interactions in the vicinity of high-affinity binding sites. A final class of sites showed enhanced cleavage by DNase I in the presence of drug. These may arise from structural distortion of the DNA helix near drug binding sites. From the point of view of the DNA sequence preference of daunorubicin binding, the unprotected and simple protected sites are of the most interest and will be discussed here.

Table 3 lists the sites whose rate of cleavage by DNase I is unaffected by the addition of daunorubicin. The information content of these aligned sequences may be evaluated by use of an algorithm presented by Stormo.[106] Such an analysis reveals that the aligned sequences in Table 3 are not entirely random and that certain positions contain a modest information content. In particular, these data reveal that cleavage of sequences to the 5' side of contiguous AT base pairs tends to be unaffected by the addition of daunorubicin. Thus, the sequence 5'(A/T)(A/T)N, where (A/T) means either A or T and N means any nucleotide, does not appear to bind daunorubicin. Macroscopic binding studies have shown that daunorubicin binds poorly to the synthetic polynucleotide poly(dA)·poly(dT),[89] a finding that is consistent with this conclusion from footprinting results.

Table 3. Analysis of Sequences Surrounding Sites Whose Rate of DNase I Digestion Was Unaffected by the Addition of Daunorubicin[a]

5'	POSITION					3'
	−3	−2	−1	+1	+2	+3
A	0.318	0.409	0.227	0.409	0.455	0.316
T	0.136	0.136	0.318	0.409	0.273	0.227
G	0.227	0.273	0.227	0.091	0.227	0.273
C	0.318	0.182	0.227	0.091	0.045	0.182
I_{sq}	0.07	0.12	0.01	0.30	0.27	0.02

Note: [a] The table shows the frequency of bases surrounding the unprotected sites, along with the calculated information content (I_{sq}) obtained by use of the algorithm of Stormo.[106] The number of sites analyzed was 22. I_{sq} ranges from 0 (no information) to 2.0 (maximum information).

Table 4. Sites and Surrounding Sequences Showing
the Greatest Protection Upon Addition of
Daunorubicin[a]

POSITION	SEQUENCE
67	5' CACTT^TACAG
70	5' TTTAC^AGCGG
59	5' AA*ACG*^TAACA
119	5' GA*CGA*^GGCCA
95	5' GA*TGC*^GCCCC
54	5' TTTCT^CAACG
36	5' TT*ACG*^CAACC
38	5' AC*GCA*^ACCAG
100	5' GCCCC^*GCT*TC
64	5' TAACA^CTTTA

Note: [a]The symbol (^) shows the protected bond in each sequence. The common
triplet sequence is italicized.

In contrast to these sites unaffected by daunorubicin, several
sequences show striking protection from DNase I digestion in
the presence of daunorubicin. The best-protected sites, revealed
by quantitative microdensitometry of autoradiograms used to
record the footprinting experiment, are shown in Table 4. These
sites appear to be completely saturated at a daunorubicin concentration of ≤ 0.25 µM. Inspection of these sites reveals that
the majority of these sequences contain a common triplet sequence, 5'(A/T)CG or 5'(A/T)GC, i.e., an AT base pair flanked
by contiguous GC base pairs in an alternating purine–pyrimidine
(or pyrimidine–purine) configuration. Because macroscopic
binding studies, and X-ray crystallographic structures, have
shown that daunorubicin binds to a 3-bp site, it is reasonable
to assume that its preferred binding sequence is a triplet. The
results of these footprinting studies are fully consistent with
the prediction of a preferred triplet daunorubicin binding site,
5'TCG, made by Pullman and co-workers,[59–63] based on their
computational studies. This is one of the few cases in which
computational chemistry has successfully predicted the behavior

of a DNA–ligand interaction that has been subsequently proven correct by experiment.[63]

Although daunorubicin most certainly binds preferentially to these triplet sequences, it must be emphasized that its binding specificity is not absolute and that the drug can and does bind to other triplet sequences, albeit with lower affinity. Footprinting titration studies allow for lower limit estimates to be made of individual site binding constants at each position where protection is registered. Figure 4 shows the extremes of behavior ob-

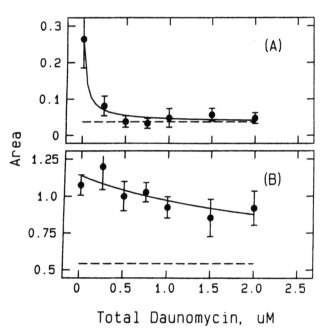

Figure 4. Individual site-binding isotherms for the interaction of daunorubicin at two specific sequences within the *tyr T* fragment. Panel A shows the highest affinity site observed in footprinting titration experiments; panel B shows the lowest affinity site that is protected. Nonlinear least-squares analysis yields a binding constant of 24.4×10^6 M^{-1} for the data in panel A and 0.4×10^6 M^{-1} for the data in panel B (values that correspond to a difference in binding free energy between the two sites of about 2 kcal mol^{-1}).

served for daunorubicin binding to the *tyr T* fragment. Binding constants at the two sites shown vary by a factor of 60, corresponding to a difference in binding free energy of about 2 kcal mol^{-1}. Figure 5 shows the distribution of experimentally determined binding constants for all of the protected sites observed in the footprinting titration experiments. The preferred sites listed in Table 4 are the high-affinity sites, and their binding constants differ from the mean value by about a factor of 5. Thus daunorubicin discriminates strongly *against* certain sequences as binding sites, the 5'(A/T)A motif, but will bind to others with a range of binding affinities.

To test the results emerging from footprinting experiments, DNA oligonucleotides were designed and synthesized for use

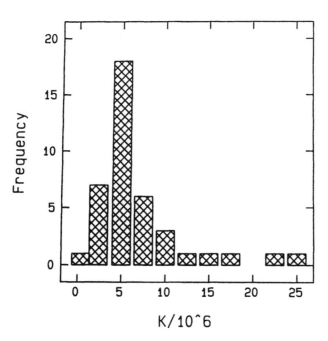

Figure 5. Distribution of the binding constants observed for the 40 protected sites monitored in footprinting studies. The mean for this distribution is 5.8 (±5.2) × 10^6 M^{-1}.

Table 5. Binding Constants for the Interaction
of Daunorubicin with Hexameric Hairpin
Molecules or Duplex Hexamers

SEQUENCE	$K_i (10^5 M^{-1})$
5'-CGTACG^{T}T GCATGC$_T$T	40.0 (±3.5)[a]
5'-CGCGCG^{T}T GCGCGC$_T$T	20.0 (±1.8)[a]
5'-TATATA^{T}T ATATAT$_T$T	11.0 (±1.0)[a]
5'-CGTACG GCATGC	4.6 (±0.4)[b]
5'-CGCGCG GCGCGC	2.3 (±0.6)[b]

Notes: [a]Satyanarayana, S., Chaires, J. B., unpublished data. Solution conditions: 20 °C, pH 7.0, 0.2 M NaCl.
[b]Data taken from Ref. 107. Solution conditions: 20 °C, pH 7.0, 1 M NaCl.

in fluorescence titration binding studies. Hexanucleotide sequences were used, one containing the putative preferred triplet sequence (5'ACG) and others of simpler repeating CG or TA sequences to serve as controls. The first study of this type was reported by Rizzo and co-workers[107] and utilized hexameric duplex DNA samples. We have reproduced their findings (S. Satyanarayana, J. B. Chaires, unpublished data) but have in addition synthesized hairpin molecules for binding studies to avoid the necessity of conducting binding studies in 1 M NaCl, a salt concentration required to stabilize the DNA hexamers. The results of studies conducted with the hairpin molecules under more standard conditions (0.2 M NaCl) are shown in Table 5. Binding is tighter to the hairpin sequence containing the triplet 5'ACG than to either of the two control hairpins. This finding is consistent with the published work utilizing hexameric duplexes[107] and strongly supports the conclusions from our footprinting titration studies.

VII. MOLECULAR DETERMINANTS OF THE DAUNORUBICIN SEQUENCE AND STRUCTURAL PREFERENCES

The preference of daunorubicin binding to the triplets 5'(A/T)CG and 5'(A/T)GC may be rationalized by use of the extensive structural[32–46] and computational[59–65] data available. First, daunorubicin is unique in that it binds to DNA by a *mixed* mode. The anthraquinone ring *intercalates* into DNA, with its long axis nearly perpendicular to the long axes of the base pairs constituting the intercalation site. The daunosamine and C-13 and C-14 portions of daunorubicin lie in the minor groove, providing *groove-binding* interactions. The key determinants of the sequence preference of daunorubicin binding appear to be the O-9 hydroxyl and, surprisingly, the daunosamine. The O-9 hydroxyl forms two hydrogen bonds with the central guanine in the preferred triplet sequence. Only a single hydrogen bond could be formed if an AT base pair were to occupy that position. The AT base pair at the 5' end of the triplet is preferred because it allows for a better stereochemical fit of the daunosamine in the minor groove. If a GC base pair were at this position, the N-2 of guanine would protrude into the minor groove and hinder the fit of the daunosamine. In addition, the N-3 amino group in daunosamine can form a direct hydrogen bond with thymine at the 5' end of the triplet. These interactions can explain the sequence preference emerging from the footprinting studies and the somewhat unusual mixed triplet motif of the preferred site. The computational, structural, and solution studies thus provide a mutually consistent picture of daunorubicin's sequence preference and its underlying molecular basis.

The strong preference of daunorubicin for B-form DNA can be rationalized as arising from interactions within the minor groove. The orientations of the A ring and its constituents and the daunosamine relative to the intercalating chromophore are such that they match almost exactly the angle made by the minor groove relative to the DNA helical axis. These two moieties fit snugly into the minor groove in a classic lock-and-key manner. Structural alterations in DNA that disrupt the normal minor groove geometry, such as conversion to the Z form or bending, accordingly offer a worse stereochemical fit for the bound drug.

VIII. BIOLOGICAL IMPLICATIONS OF DAUNORUBICIN BINDING TO PREFERRED SITES

A recent study[108] showed that doxorubicin, in the absence of excision repair, induced highly specific deletion and base-substitution mutations in *Escherichia coli*. A striking finding was that 80% of the deletion mutation was found at the end of the triplet sequences 5′(A/T)CG or 5′(A/T)GC, the very triplets identified as preferred binding sites in our footprinting studies. Thus, the sequence preference inferred in vitro persists in vivo.

The existence of a *distribution* of binding sites along the DNA lattice has several implications. First, it has been argued[9] that the magnitude of observed macroscopic DNA binding constants ($\approx 10^6$ M^{-1}) is too low to account for the physiological concentrations ($\approx 10^{-8}$ M) at which the anthracycline antibiotics exert their effects and that something other than DNA alone must, therefore, be the key target of the drugs. This view must be reconsidered in light of our studies. Our footprinting results show that selected DNA sites can bind daunorubicin with binding constants at least an order of magnitude greater than the overall macroscopic binding constant. These sites will generally not be observed in macroscopic binding studies owing to their low frequency and to the inherent difficulty in measuring tight binding by the optical methods typically employed in such studies. High-affinity sites with binding constants of greater than 10^7 M^{-1} exist on DNA and would be selectively occupied under "physiological" drug concentrations.

Second, it is common to attempt to correlate the inhibition of some biological activity (DNA replication, topoisomerase or helicase activity, etc.) with some macroscopic measure of binding, i.e., K or ΔT_m values. Because binding to DNA is best described by a distribution of binding constants, such attempts are likely to be unsatisfactory and will neglect the possible preferential binding of antibiotic to certain sequences. Because many biological activities utilizing the DNA template rely on specific regulatory sequences, a better correlation between anthracycline binding to DNA and these activities might be made by recognition of the broad distribution of affinities along the DNA lattice and by use of the appropriate microscopic drug-binding constant.

ACKNOWLEDGMENTS

Work in the author's laboratory has been generously supported by Grant CA35635 from the National Cancer Institute. Dr. Julio Herrera participated in quantitative footprinting studies, and his substantial contributions are gratefully acknowledged.

REFERENCES

1. Weiss, R. *Sem. in Oncology* **1992**, *19*, 670–686.
2. Fukushima, M. *Nature* **1989**, *342*, 850–851.
3. Arcamone, F. *Doxorubicin;* Academic Press: New York, 1981.
4. *Anthracycline and Anthracenedione Based Anticancer Agents;* Lown, J. W., Ed.; Elsevier: Amsterdam, 1988.
5. *Anthracyclines: Current Status and New Developments;* Crooke, S. T., Reich, S. D., Eds.; Academic Press: New York, 1982.
6. *Anthracycline Antibiotics: New Discoveries, Methods of Discovery, and Mechanism of Action;* Priebe, W., Ed.; ACS Symposium Series 574, American Chemical Society: Washington, D.C., 1994.
7. Lown, J. W. *Chem. Soc. Rev.* **1993**, *22*, 165–176.
8. Pullman, B. *Adv. Drug Res.* **1989**, *18*, 1–113.
9. Chabner, B. A.; Collins, J. M. *Cancer Chemotherapy: Principles and Practice;* J. B. Lippincott Co.: Philadelphia, 1990, Chapter 14.
10. Chaires, J. B. *Biophys. Chem.* **1990**, *35*, 191–202.
11. Fritzche, H.; Walter, A. In *Chemistry—Physics of DNA–Ligand Interactions*; Kallenbach, N. R., Ed.; Adenine Press: Schenectady, 1989, pp 1–35.
12. Arcamone, F. In *Molecular Aspects of Chemotherapy;* Borowski, E., Shugar, D., Eds.; Pergamon Press: New York, 1989, pp 47–53.
13. Fritzche, H.; Berg, H. *Gazz. Chim. Ital.* **1987**, *117*, 331–352.
14. Abdella, B. R. J.; Fisher, J. *Environ. Health Perspectives* **1985**, *64*, 3–18.
15. Aubel-Sadron, G.; Londos-Gagliardi, D. *Biochimie* **1984**, *66*, 333–352.
16. Brown, J. R.; Imam, S. H. *Prog. Med. Chem.* **1984**, *21*, 169–236.
17. Gianni, L.; Corden, B. J.; Myers, C. F. *Rev. Biochem. Toxicol.* **1983**, *5*, 1–82.
18. Brown, J. R. *Prog. Med. Chem.* **1978**, *15*, 125–164.
19. Porumb, H. *Prog. Biophys. Molec. Biol.* **1978**, *34*, 175–195.

20. Neidle. S. In Sammes, P. G. (ed.) *Topics in Antibiotic Chemistry,* Vol. 2., Halsted Press, Chichester, 1978, pp 241–278.
21. DiMarco, A.; Arcamone, F.; Zunino, F. In *Antibiotics III: Mechanism of Action of Antimicrobial and Antitumor Agents;* Corcoran, J. W., Hahn, F. E., Eds.; Springer-Verlag: New York, 1975, pp 101–128.
22. Tritton, T. R. *Pharmacol. Ther.* **1991,** *49,* 293–309.
23. Keizer, H. G.; Pinedo, H. M.; Schuurhuis, G. J.; Joenje, H. *Pharmacol. Ther.* **1990,** *47,* 219–231.
24. Valentini, L.; Nicolella, V.; Vannini, E.; Menozzi, M.; Penco, S.; Arcamone, F. *Il Farmaco* **1985,** *40,* 377–390.
25. Gigli, M.; Doglia, S. M.; Millot, J. M.; Valentinin, L; Manfait, M. *Biochim. Biophys. Acta* **1988,** *950,* 13–20.
26. Belloc, F.; Lacombe, F.; Dumain, P.; Lopez, F.; Bernard, P.; Boisseau, M. R.; Reifers, J. *Cytometry* **1992,** *13,* 880–885.
27. Nabiev, I.; Sokolov, K.; Morjani, H.; Manfait, M. In *Spectroscopy of Biological Molecules;* Hester, R. E., Girling, R. B., Eds.; The Royal Society of Chemistry: Cambridge, U.K., 1991, pp 345–348.
28. Bartkowiak, J.; Kapuscinski, J.; Melamed, M. R.; Darzynkiewicz, Z. *Proc. Natl. Acad. Sci. U.S.A.* **1989,** *86,* 5151–5154.
29. Liu, L. F. *Ann. Rev. Biochem.* **1989,** *58,* 351–375.
30. Bodley, L.; Liu, L. F.; Israel, M.; Seshadri, R.; Koseki, Y.; Giuliani, F. C.; Kirschenbaum, S.; Silber, R.; Potmesil, M. *Cancer Res.* **1989,** *49,* 5969–5978.
31. Capranico, G.; Zunino, F.; Kohn, K.; Pommier, Y. *Biochemistry* **1990,** *29,* 562–569.
32. Quigley, G. J.; Wang, A. H.-J.; Ughetto, G.; van der Marel, G.; van Boom, J. H.; Rich, A. *Proc. Natl. Acad. Sci. U.S.A.* **1980,** *77,* 7204–7208.
33. Wang, A. H.-J.; Ughetto, G.; Quigley, G. J.; Rich, A. *Biochemistry* **1987,** *26,* 1152–1163.
34. Holbrook, S. R.; Wang, A. H.-J.; Rich, A.; Kim, S.-H. *J. Mol. Biol.* **1988,** *199,* 349–357.
35. Moore, M. H.; Hunter, W. N.; d'Estaintot, B. L.; Kennard, O. *J. Mol. Biol.* **1989,** *206,* 693–705.
36. Frederick, C. A.; Williams, L. D.; Ughetto, G.; van Boom, J. H.; Rich, A.; Wang, A. H.-J. *Biochemistry* **1990,** *29,* 2538–2549.
37. Nunn, C. M.; van Meervelt, L.; Zhang, S.; Moore, M. H.; Kennard, O. *J. Mol. Biol.* **1991,** *222,* 167–177.
38. Wang, A. H.-J.; Gao, Y.-G.; Liaw, Y.-C.; Li, Y.-K. *Biochemistry* **1991,** *30,* 3812–3815.
39. Leonard, G. A.; Hambly, T. W.; McAuley-Hecht, K; Brown, T.; Hunter, W. N. *Acta Cryst.* **1993,** *D49,* 458–467.
40. Limpscomb, L. A.; Peek, M. E.; Zhou, F. X.; Bertrand, J. A.; VanDerveer, D.; Williams, L. D. *Biochemistry* **1994,** *33,* 3649–3659.
41. Williams, L. D.; Frederick, C. A.; Ughetto, G.; Rich, A. *Nucleic Acids Res.* **1990,** *18,* 5533–5541.
42. Leonard, G. A.; Brown, T.; Hunter, W. N. *Eur. J. Biochem.* **1992,** *204,* 69–74.

43. d'Estaintot, B. L.; Gallois, B.; Brown, T.; Hunter, W. N. *Nucleic Acids Res.* **1992**, *20*, 3561–3566.
44. Gao, Y.-G.; Wang, A. H.-J. *Anti-Cancer Drug Design* **1991**, *6*, 137–149.
45. Gallois, B.; d'Estaintot, B. L.; Brown, T.; Hunter, W. N. *Acta Cryst.* **1993**, *D49*, 311–317.
46. Cirilli, M.; Bachechi, F.; Ughetto, G.; Colonna, F. P.; Capobianco, M. L. *J. Mol. Biol.* **1992**, *230*, 878–889.
47. Patel, D. J.; Canuel, L. L. *Eur. J. Biochem.* **1978**, *90*, 247–254.
48. Patel, D. J. *Biopolymers* **1979**, *18*, 553–569.
49. Phillips, D. R.; Roberts, G. C. K. *Biochemistry* **1980**, *19*, 4795–4801.
50. Patel, D. J.; Kozlowski, S. A.; Rice, J. A. *Proc. Natl. Acad. Sci. U.S.A.* **1981**, *78*, 3333–3337.
51. Tran-Dinh, S.; Cavailles, J.-A.; Herve, M.; Neuman, J.-M.; Garnier, A.; Huyn-Dihn, T.; d'Estaintot, B. L.; Igolen, J. *Nucleic Acids Res.* **1984**, *12*, 6259–6278.
52. Ragg, E.; Battistini, C.; Garbesi, A.; Colonna, F. P. *FEBS Lett.* **1988**, *236*, 231–234.
53. Hammer, B. C.; Russell, R. A.; Warrener, R. N.; Collins, J. G. *Eur. J. Biochem.* **1989**, *191*, 683–688.
54. Gorenstein, D. G.; Lai, K. *Biochemistry* **1989**, *28*, 2804–2812.
55. Odefey, C.; Westendorf, J.; Dieckmann, T.; Oschkinat, H. *Chem.–Biol. Interactions* **1992**, *85*, 117–126.
56. Manfait, M.; Alix, A. J. P.; Jeannesson, P.; Jardillier, J.-C.; Theophanides, T. *Nucleic Acids Res.* **1982**, *10*, 3803–3816.
57. Smulevich, G.; Feis, A. *J. Phys. Chem.* **1986**, *90*, 6388–6392.
58. Pohle, W.; Bohl, M.; Flemming, J.; Bohlig, H. *Biophys. Chem.* **1990**, *35*, 213–226.
59. Chen, K.-X.; Gresh, N.; Pullman, B. *J. Biomolec. Struct. Dynam.* **1985**, *3*, 445–466.
60. Chen, K.-X.; Gresh, N.; Pullman, B. *Nucleic Acids Res.* **1986**, *14*, 2251–2267.
61. Chen, K.-X.; Gresh, N.; Pullman, B. *Mol. Pharmacol.* **1986**, *30*, 279–286.
62. Gresh, N.; Pullman, B.; Arcamone, F.; Menozzi, M.; Tonani, R. *Mol. Pharmacol.* **1989**, *35*, 251–256.
63. Pullman, B. *Anti-Cancer Drug Design* **1990**, *7*, 95–105.
64. Cieplak, P.; Rao, S. N.; Grootenhuis, P. D. J.; Kollman, P. *Biopolymers* **1990**, *29*, 717–727.
65. Chaires, J. B.; Dattagupta, N. D.; Crothers, D. M. *Biochemistry* **1982**, *21*, 3933–3940.
66. Chaires, J. B. *Biopolymers* **1985**, *24*, 403–419.
67. Graves, D. E.; Krugh, T. R. *Biochemistry* **1983**, *22*, 3941–3947.
68. Barcelo, F.; Martorell, J.; Gavilanes, F.; Gonzalez-Ros, J. M. *Biochem. Pharmacol.* **1988**, *37*, 2133–2138.
69. Schutz, H.; Gollmick, F. A.; Stutter, E. *Studia Biophysica* **1979**, *75*, 147–159.
70. McGhee, J. D.; von Hippel, P. H. *J. Mol. Biol.* **1974**, *86*, 469–489.
71. Remeta, D. P.; Mudd, C. P.; Berger, R. L.; Breslauer, K. J. *Biochemistry* **1991**, *30*, 9799–9809.

72. Remeta, D. P.; Mudd, C. P.; Berger, R. L.; Breslauer, K. J. *Biochemistry* **1993**, *32*, 5064–5073.

73. Chaires, J. B.; Priebe, W.; Graves, D. E.; Burke, T. G. *J. Am. Chem. Soc.* **1993**, *115*, 5360–5364.

74. Chaires, J. B. In *Advances in DNA Sequence Specific Agents*, Vol. 1; Hurley, L. H., Ed.; JAI Press, Inc.: Greenwich, 1992, pp 3–23.

75. Stutter, E.; Forster, W. *Studia Biophysica* **1979**, *75*, 199–208.

76. Forster, W.; Stutter, E. *Int. J. Biol. Macromol.* **1984**, *6*, 114–124.

77. Chaires, J. B.; Dattagupta. N.; Crothers, D. M. *Biochemistry* **1985**, *24*, 260–267.

78. Krishnamoorthy, C. R.; Yen, S.-F.; Smith, J. C.; Lown, J. W.; Wilson, W. D. *Biochemistry* **1986**, *25*, 5933–5940.

79. Phillips, D. R.; Greif, P. C.; Boston, R. C. *Mol. Pharmacol.* **1988**, *33*, 225–230.

80. Rizzo, V.; Sacchi, N., Menozzi, M. *Biochemistry* **1989**, *28*, 274–282.

81. Chaires, J. B. *Nucleic Acids Res.* **1983**, *11*, 8485–8494.

82. Van Helden, P. D. *Nucleic Acids Res.* **1983**, *11*, 8415–8420.

83. Chen, C.-W.; Knop, R. H.; Cohen, J. S. *Biochemistry* **1983**, *22*, 5468–5471.

84. Chaires, J. B. *Biochemistry* **1985**, *24*, 7479–7486.

85. Britt, M.; Zunino, F.; Chaires, J. B. *Mol. Pharmacol.* **1986**, *29*, 74–80.

86. Chaires, J. B. *J. Biol. Chem.* **1986**, *261*, 8899–8907.

87. Chaires, J. B. *Biochemistry* **1986**, *25*, 8436–8439.

88. Neumann, J. M.; Cavailles, J. A.; Herve, M.; Tran-Dinh, S.; d'Estaintot, B. L.; Huynh-Dinh, T.; Igolen, J. *FEBS Lett.* **1985**, *182*, 360–364.

89. Herrera, J. E.; Chaires, J. B. *Biochemistry* **1989**, *28*, 1993–2000.

90. Chaires, J. B.; Dattagupta, N.; Crothers, D. M. *Biochemistry* **1983**, *22*, 284–292.

91. Fritzsche, H.; Wahnert, U.; Chaires, J. B.; Dattagupta, N.; Schlessinger, F. B.; Crothers, D. M. *Biochemistry* **1987**, *26*, 1996–2000.

92. Portugal, J.; Waring, M. J. *Biochemie* **1987**, *69*, 825–840.

93. Cera, C.; Palu, G.; Magno, S. M.; Palumbo, M. *Nucleic Acids Res.* **1991**, *19*, 2309–2314.

94. Van Dyke, M. W.; Hertzberg, R. P.; Dervan, P. B. *Proc. Natl. Acad. Sci. U.S.A.* **1982**, *79*, 5470–5474.

95. Lane, M. J.; Dabrowiak, J. C.; Vournakis, J. N. *Proc. Natl. Acad. Sci. U.S.A.* **1983**, *80*, 3260–3264.

96. Low, C. M. L.; Drew, H. R.; Waring, M. J. *Nucleic Acids Res.* **1984**, *12*, 4865–4879.

97. Portugal, J. *Chem.–Biol. Interactions* **1989**, *71*, 311–324.

98. Goodisman, J.; Dabrowiak, J. C. In *Advances in DNA Sequence Specific Agents*, Vol. 1; Hurley, L. H. , Ed.; JAI Press, Inc.: Greenwich, 1992, pp 25–49.

99. Chaires, J. B.; Fox, K. R.; Herrera, J. E.; Britt, M.; Waring, M. J. *Biochemistry* **1987**, *26*, 8227–8236.

100. Skorobogaty, A.; White, R. J.; Phillips, D. R.; Reiss, J. A. *FEBS Lett.* **1988**,*227*, 103–106.

101. Skorobogaty, A.; White, R. J.; Phillips, D. R.; Reiss, J. A. *Drug Design and Delivery* **1988**, *3*, 125–152.

102. Chaires, J. B.; Herrera, J. E.; Waring, M. J. *Biochemistry* **1990**, *29*, 6145–6153.

103. Chaires, J. B. In *Molecular Basis of Specificity in Nucleic Acid–Drug Interactions;* Pullman, B., Jortner, J., Eds.; Kluwer Academic Press: Dordrecht, The Netherlands, 1990, pp 123–136.
104. Phillips, D. R.; Crothers, D. M. *Biochemistry* **1986**, *25*, 7355–7362.
105. Phillips, D. R.; Cullinane, C.; Trist, H.; White, R. J. In *Molecular Basis of Specificity in Nucleic Acid–Drug Interactions;* Pullman, B., Jortner, J., Eds.; Kluwer Academic Press: Dordrecht, The Netherlands, 1990, pp 137–155.
106. Stormo, G. D. *Meth. Enzymol.* **1991**, *208*, 458–468.
107. Rizzo, V.; Battistini, C.; Vigevani, A.; Sacchi, N.; Razzano, G.; Arcamone, F.; Garbesi, A.; Colonna, F. P.; Capobianco, M.; Tondelli, L. *J. Mol. Recog.* **1989**, *2*, 132–141.
108. Anderson, R. D.; Veigl, M. L.; Baxter, J.; Sedwick, W. D. *Cancer Res.* **1991**, *51*, 3930–3937.

COVALENT INTERACTIONS OF ETHIDIUM AND ACTINOMYCIN D WITH NUCLEIC ACIDS:

PHOTOAFFINITY LABELING OF DNA

David E. Graves

Advances in DNA Sequence Specific Agents
Volume 2, pages 169–186.
Copyright © 1996 by JAI Press Inc.
All rights of reproduction in any form reserved.
ISBN: 1-55938-166-3

I. INTRODUCTION

The mechanism through which small molecules such as chemotherapeutic agents, mutagens, and carcinogens interact with nucleic acids is of central biological significance. The structural as well as thermodynamic and kinetic parameters associated with the ligand–DNA complexes serve as a basis of our current knowledge of many biologically active compounds and are used to bridge the physicochemical properties associated with formation of these complexes with various biological effects that are induced by the compounds, such as inhibition of DNA replication and transcription, topoisomerase II inhibition, mutagenesis, and carcinogenesis.[1]

Over the past decade, the determination of the sequence selective interaction of small molecules with nucleic acids has emerged as a key problem in the field of ligand–DNA interactions.[2] Central to our characterization of drug–DNA interactions is the development of a basic understanding of the underlying properties that dictate the physical and chemical nature of sequence selective binding of ligands to DNA. However, even with recent advances in areas such as quantitative DNA footprinting, relatively little information has been obtained to explain the molecular basis of sequence specificity. One approach toward gaining insight into the mechanism through which DNA binding ligands target specific sequences is the technique of photoaffinity labeling.

The usefulness of photoreactive agents as tools for probing structural and functional properties of the active sites of proteins has been well documented for more than two decades; however, relatively few applications of photoaffinity labeling have been applied toward the study of ligand–DNA interactions.[3–5] Yielding and co-workers realized the usefulness of this method in the early 1970s and successfully developed ethidium monoazide

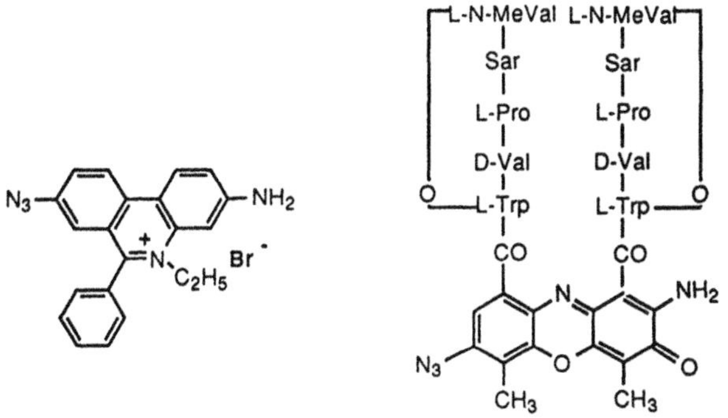

Figure 1. Chemical structure of the photoreactive analogues of actinomycin D (7-azidoactinomycin D) and ethidium bromide (8-azidoethidium bromide).

(8-azido-3-amino-6-phenyl-5-ethylphenanthradinium) bromide (Figure 1) as a tool for examining ligand–DNA interactions.[6,7] Their studies demonstrated the utility of this method for the characterization of ligand–DNA interactions and the use of these drugs as probes for examining mitochondrial mutagenesis, frameshift mutagenesis, and DNA repair.[8–12]

Earlier studies using a variety of thermodynamic, spectroscopic, and kinetic measurements revealed that placement of the azido moiety on the phenanthridine ring (ethidium monoazide) or phenoxazone ring (in the case of 7-azidoactinomycin D) does not significantly alter the intercalative binding properties of these photoaffinity analogues to DNA.[13,14] Both the ethidium and actinomycin D photoaffinity probes exhibit DNA-binding affinities, binding site sizes, and base sequence specificities (in the absence of light) equivalent to those of their parent compounds. However, upon photolysis by visible light, the azide moiety is converted to a highly reactive nitrene that forms a covalent bond in situ.[15] We have exploited this ability for covalent attachment of ethidium or actinomycin D to DNA to probe DNA-binding specificities at markedly lower drug concentrations than had been possible for studies involving the reversibly binding parent molecules, thereby providing much greater sensitivity for characterizing the sequence selective binding of ligands to DNA.[16]

Similarly, photoaffinity labeling with ethidium monoazide was successfully used to probe the energetic and kinetic properties associated with structural changes of DNA (i.e., B to Z transitions). Salt-induced structural changes of poly(dG–dC)·poly(dG–dC) were shown to be highly influenced by covalent attachment of ethidium to the DNA. The ethidium–DNA adduct was formed under B-DNA conditions, effectively locking the intercalation site in a right-hand conformation. Addition of salt (conducive for the B to Z conformational change) was monitored by absorption and CD spectroscopies. The presence of covalently attached ethidium to the DNA lattice was highly influential in inhibiting the B to Z transition. Effects of temperature, ionic strengths, and drug binding densities were used to examine the energetics, kinetics, and mechanistic properties of the structural transformations between the B and Z conformations.[17,18]

Actinomycin D has long been the paradigm of a sequence specific DNA binding agent. As early as 1968, actinomycin D was shown to demonstrate a specificity for guanine, and an intercalation model in 1972 by Jain and Sobell predicted that the dGpC step would be a high-affinity site for actinomycin D binding.[19,20] The recent development of the photoreactive analogue of actinomycin D has allowed us to probe directly the high-affinity binding sites of actinomycin D at very low levels of bound drug. Studies by Rill and co-workers[16] using piperidine cleavage and DNA-sequencing methods have demonstrated several novel atypical (non-dGpC) actinomycin D binding sites; most notable among these sites is the dTG_nT motif.[16,21] These studies demonstrate the existence of additional binding sites along with information concerning the marked influenced of the flanking base sequences adjacent to either side of the intercalation site on the sequence specificity and actinomycin D binding.[22] The ability for covalent attachment of actinomycin D in a rapid manner has recently been used to probe the mechanism through which actinomycin D exerts DNA sequence specificity. Over a decade ago, analyses of complex association kinetics data led Fox and Waring to propose the "shuffling hypothesis" whereby actinomycin D binds to the DNA lattice in a relatively nonspecific manner and "shuffles" along the helix in search of high-affinity binding sites.[23,24] Recently, Waring and co-workers utilized 7-azidoactinomycin D to covalently attach the drug to the DNA lattice as a function of equilibration time, followed by DNA-sequencing analysis to determine the site of attachment of the drug. Results of these studies clearly indicate that initial binding of actinomycin D to the DNA is relatively non–sequence specific in GC-rich regions of the DNA. However, over a period of time, the drug is demonstrated to decrease in binding density from certain sites and increase in binding density at more favorable binding sites, supportive of the shuffling hypothesis.[24,25]

II. PHOTOAFFINITY LABELING TO PROBE ETHIDIUM–DNA INTERACTIONS

Ethidium bromide is one of the most valuable and intensively studied of the DNA-binding ligands. It has been used to examine the structural and functional properties of DNA and as a model compound for characterizing intercalative binding processes for over three decades.[26,27] However, the use of ethidium is limited owing to the reversible nature of the ethidium–DNA interaction. Photoaffinity labeling, a technique that involves the design and synthesis of analogues with the ability to covalently attach at their binding sites when photoactivated, provides a means of overcoming this limitation. Photoreactive analogues of ethidium have been synthesized by replacement of one or both of the amino groups at the 3- or 8-positions on the phenanthidine ring with an azido substituent, and these analogues have been used to provide an initial characterization of the ethidium–DNA adduct as a model system for studying frameshift mutagenesis and an inducer of DNA repair processes.[9–13,28,29]

The usefulness of the 8-azido compound as a photoaffinity analogue is reflected in its ability to mimic the noncovalent DNA interaction of the parent ethidium bromide. Comparative studies of the binding of ethidium and its photoreactive analogues to DNA indicate that the interaction of the 3-amino-8-azido derivative (ethidium monoazide) with DNA is identical

Table 1. Equilibrium Binding Properties of Ethidium and
Actinomycin D Photoaffinity Analogues

Property	Ethidium Bromide (Parent Compound)		Ethidium Monoazide[a]	
Absorption max (nm)	480 (free)	520 (bound)	456 (free)	495 (bound)
K_{int} $(M^{-1})^b$	5 × 10⁵		4 × 10⁵	
n (base pairs)	2		2	
	Actinomycin D (Parent Compound)		7-Azidoactinomycin D	
Absorption max (nm)	440 (free)	452 (bound)	462 (free)	472 (bound)
K_{int} $(M^{-1})^b$	5 × 10⁵		5 × 10⁵	
n (base pairs)[c]	7		7	

Notes: [a]3-Amino-8-azido-6-phenyl-5-ethylphenanthdinium bromide.
[b]K_{int} is the intrinsic DNA binding affinity determined by use of the neighbor exclusion model.
[c]*n* is the binding site size in base pairs (i.e. base pairs per bound drug) at saturation.[16]

to that of the parent compound, as shown by the data in Table 1.[14]

Ethidium bromide binds to DNA via an intercalation mechanism in which the planar aromatic phenanthridinium ring is inserted between adjacent base pairs without disrupting the Watson–Crick hydrogen bonding. In 1980, Yielding and co-workers[30] revealed that the reactions of the monoazide and ethidium with DNA resulted in relatively large increases in the fluorescence emission signal upon binding with DNA, suggesting that the monoazide intercalates in a manner similar to ethidium. From stopped-flow studies, it was found that the parent ethidium and photoreactive ethidium azide exhibit identical apparent first-order rate constants upon binding to DNA.[30,31] Similarly, the association constants and site exclusion sizes for the two compounds are identical,[10] demonstrating the usefulness of this photoreactive probe to characterize ethidium–DNA interactions.

In early studies, Yielding and co-workers used the photoreactive analogue of ethidium to probe various in vivo processes such as mitochondrial mutagenesis, frameshift mutagenesis, and DNA repair. These studies revealed that addition and subsequent photolysis of ethidium monoazide in yeast cells results in enhanced production of petite mutations when compared to similar concentrations of reversibly binding ethidium bromide.[9] Similarly, ethidium bromide was shown to be nonmutagenic in the *Salmonella typhimurium* strain TA1578, but in contrast, photolysis of bacteria incubated with ethidium monoazide resulted in the observation of frameshift mutagenesis in the repair-deficient bacterial strains. The repair-proficient strains were shown to be protected from this mutagenesis, indicating that the frameshift mutation was due to covalent adduct formation between the DNA and photoaffinity analogue of ethidium.[12,13] DNA repair processes in eukaryotic cells (human lymphocytes) were also probed with ethidium monoazide. These studies demonstrated that drug concentrations as low as 1 µM could provoke considerable DNA repair synthesis.[28,32] The ability to simultaneously measure both the extent and distribution of drug excision along with incorporation of DNA bases provided a novel and highly efficient approach for characterization of the mechanistic properties associated with DNA repair.

III. SEQUENCE SPECIFICITY OF ETHIDIUM–DNA INTERACTIONS

Because of the reversible nature of the ethidium–DNA interaction, relatively high concentrations of drug are required to probe the base sequence specificity of ethidium with DNA. However, as concentrations of the drug are increased, the ability to detect high-affinity binding sites is diminished, which led early investigators to conclude a lack of base sequence specificity for ethidium. In contrast, the use of 8-azidoethidium allows covalent attachment of the drug to the DNA at very low concentrations and thus provides a much more sensitive measure of sequence specificity. In 1982, Yielding and co-workers used this novel approach to demonstrate high-affinity ethidium binding sites on plasmid DNA.[33] In this study, low levels of ethidium were covalently attached to the plasmid pBR322, followed by digestion with the restriction endonuclease *HhaI*

[d(GCGC)]. Their studies demonstrated that very low drug
binding densities (<1 drug molecule per 400 base pairs) resulted in *Hha*I resistance over a wide range of enzyme concentrations. *Hha*I digestion revealed the blockage to be highly
selective, requiring as little as one covalently attached drug
molecule per plasmid genome.

Most DNA footprinting methods rely on single-hit kinetics
from DNase I treatment of the DNA to resolve ligand binding
sites on the DNA fragments.[34,35] Rill and co-workers have demonstrated that covalent adducts formed by ethidium azide (and
7-azidoactinomycin D) are labile in the presence of piperidine
(a common sequencing reaction reagent).[16] Examination of the
consequences of photoaddition on oligonucleotide duplexes by
electrophoresis before and after treatment with hot piperidine
revealed that spontaneous photochemical cleavage of DNA did
not occur except at sites where drug was covalently attached.
From use of techniques analogous to DNA sequencing, frequencies of individual photoadduct formation for more than 1200
bases in DNA restriction fragments were determined, and surprisingly, ethidium azide, whose parent compound is thought
to be relatively nonspecific, was shown to have marked DNA
binding selectivity. Significant influences of nearest and next-
nearest neighbor bases were observed.

Covalent attachment sites were shown to be nonrandomly distributed, with the order of preference for covalent attachment
sites being G > C ≥ T > A. Analyses of neighboring base effects
revealed that guanine adducts were most reactive when flanked
on the 5′-side by guanine and on the 3′-side by either guanine
or thymine. Cytosine adducts were most reactive when flanked
on the 5′-side by cytosine and on the 3′-side by guanine.[36] The
reactivities of ethidium with the d(CpG) and d(GpG) sequences
were predicted from the optical and NMR studies of Krugh and
co-workers.[37,38] Overall analyses of all possible intercalation
sites revealed a binding order of d(CpG) ≈ d(GpG) > d(GpC)
≈ d(TpG) ≥ d(GpT) > d(GpA) ≈ d(ApG) ≈ d(TpA) > d(ApT)
> d(ApA).

IV. PROBING OF THE CONFORMATIONAL TRANSITIONS OF DNA BY PHOTOAFFINITY LABELING

Covalent attachment of ethidium to DNA results in an intercalative drug–DNA adduct that can be used to probe DNA structure. This approach was used by Gilbert and co-workers to probe the influence of intercalating drugs on the B to Z transition of poly(dG–dC)·poly(dG–dC).[17,18] In this study, the effects of adduct formation with poly(dG–dC)·poly(dG–dC) and poly(dGm⁵–dC)·poly(dGm⁵–dC) on the salt-induced B to Z transition were examined. Ethidium monoazide was allowed to incubate with DNA (in the absence of light) and form intercalation complexes with the B-DNA structure. Subsequently, the drug–DNA complexes were photolyzed, rendering the complex irreversible and essentially locking the intercalation site into a right-hand conformation. Subsequent titrations of high concentrations of salt into these drug–DNA adducts revealed that the presence of covalently bonded ethidium greatly reduced the ability of the DNA to undergo a B to Z structural transition.

A plot of the fraction of DNA in the Z conformation versus r (covalently attached ethidium per base pair) shown in Figure 2 demonstrates a linear decrease in the fraction of DNA in the Z conformation as a function of increasing amounts of covalently attached drug. The high-salt data (5 M NaCl) converge with the control low-salt data (0.1 M NaCl) at an r value of 0.4 corresponding to one covalently bound ethidium per 2.5 base pairs, indicating that covalent attachment of the ethidium monoazide effectively freezes 2 to 3 base pairs of the poly(dG–dC)·poly(dG–dC) into a right-handed conformation, presumably at the intercalation site.

Energetics of the B to Z structural transition of poly(dG–dC)·poly(dG–dC) and poly(dGm⁵–dC)·poly(dGm⁵–dC) were determined from the transition curves shown in Figure 2.[20–22] The equilibrium transition can be interpreted according to the two-state model for an intramolecular change between two states by the equation:

$$\Delta G_{app} = -RT \ln [K_a (NaCl)^n] = \Delta G^0 - nRT \ln [NaCl]$$

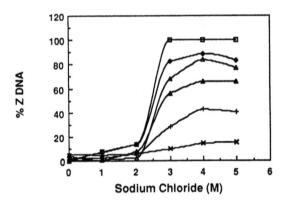

Figure 2. Curves for the salt-induced B to Z transition of poly(dG–dC)·poly(dG–dC) with the following amounts of covalently attached ethidium per base pair: □, $r = 0$; ◆, $r = 0.037$; △, $r = 0.070$; ▲, $r = 0.110$; +, $r = 0.226$; ×, $r = 0.293$. Data were obtained by circular dichroism and UV absorption spectroscopies.

Plots of the B to Z transition free energy, ΔG_{app} (determined from the above equation), as a function of ln [NaCl] are linear with a slope equal to $-nRT$ and the intercept equal to ΔG^0.

At salt conditions conducive for the Z conformation to exist, the ΔG_{app} is negative (approximately -4 kcal/mol) for both poly(dG–dC)·poly(dG–dC) and poly(dGm^5–dC)·poly(dGm^5–dC), indicative of spontaneous adoption of the Z conformation. Covalent attachment of ethidium monoazide results in more positive changes in the free energy of the B to Z transition. This effect is presumed to be through stabilization of the right-handed conformation of short regions (2 to 3 base pairs) at the intercalation site. The free energy associated with the B to Z transition for the ethidium-modified copolymers becomes positive at approximately 1 covalently bound drug per 10 base pairs, as shown in Figure 3.

These results demonstrate that covalent attachment of ethidium monoazide is highly effective in inhibiting the B to Z transition of poly(dG–dC)·poly(dG–dC) and poly(dGm^5–dC)·poly(dGm^5–dC). Intercalation of ethidium monoazide and subsequent covalent attachment in situ results in an irreversible

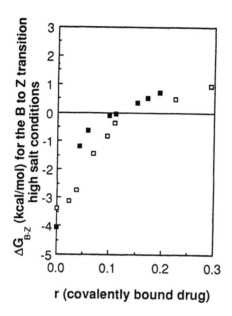

Figure 3. The free energy of the salt-induced B to Z transition is plotted as a function of covalent modification of poly(dG–dC)·poly(dG–dC) (open squares) and poly(dGm^5–dC)·poly(dGm^5–dC) (closed squares). All data were determined at 5 M NaCl for poly(dG–dC)·poly(dG–dC) and 1.5 M NaCl for poly(dGm^5–dC)·poly(dGm^5–dC), respectively. Increasing levels of modification by ethidium monoazide results in a positive increase in the free energy of the B to Z transition, making the transition less favorable. Both alternating copolymers experience a change from negative to positive ΔG_{app} at a level of 1 bound drug molecule per 10 base pairs.

right-handed nucleation site, effectively blocking the transition from the B to Z conformations. Propagation of the transition is interrupted by the right-handed binding sites, resulting in a markedly decreased cooperativity of the transition and change in the effective Na$^+$ ion binding differential between the B and Z conformations of DNA.[17]

Covalent attachment of ethidium monoazide to poly(dG–dC)·poly(dG–dC) and poly(dGm^5–dC)·poly(dGm^5–dC) was used to probe the kinetic properties associated with the B and Z transitions.[18] These studies, which compare the effects of covalently and noncovalently attached ethidium on the kinetics of the NaCl-induced B to Z transition in poly(dG–dC)·poly(dG–dC), provide considerable insight into the mechanism by which these intercalators inhibit the rate of the B to Z transition. They show that the primary effect is at the nucleation step and that inhibition of the B to Z transition by intercalators may be quantitatively correlated with a reduced probability of finding a long enough stretch of drug-free DNA suitable for nucleation.

A comparison of the relative rates of the B to Z transition with increasing binding ratios of both covalently and noncovalently bound ethidium demonstrates that both covalent and noncovalently bound ethidium exert the same effects on the rates of the B to Z transition, providing considerable insight into the underlying mechanism by which the transition is inhibited. The B to Z transition is thought to proceed by a rate-limiting nucleation step, followed by propagation from the nucleation sites.[39] Intercalation of ethidium (or covalent attachment of the intercalated ethidium monoazide) could inhibit either the nucleation or the propagation steps or both.[40] Our results are consistent with a mechanism in which the primary effect of ethidium is to reduce the probability of nucleation of the B to Z transition and thereby slow the overall rate of the transition.

The effects of covalently bonded ethidium on the reverse Z to B reaction was also examined. In this case, covalently attached ethidium should increase the rate, since the ligand should lock the helix into a right-handed conformation that would serve as a nucleation site. Such an effect is in fact observed, and the time courses of the reverse Z to B transition are adequately described by a single exponential. The rate constants for this transition as a function of covalently attached ethidium and temperature were collected for the forward and reverse (B to Z and Z to B) transitions and used to construct an Arrhenius plot to determine the influence of covalent modification on the activation energies for the B and Z transitions. Surprisingly, the activation energies for the forward (B to Z) and backward (Z to B) transitions are relatively insensitive to the presence of covalently attached drug, indicating that the influence of intercalating drugs on the structural transitions is at the nucleation rather than the propagation step.

V. 7-AZIDOACTINOMYCIN D: A PHOTOAFFINITY PROBE OF ACTINOMYCIN D

The success of ethidium monoazide as an effective photoaffinity probe for examination of ethidium–DNA interactions resulted in a search for other photoreactive DNA-binding agents. Among these, the most advantageous to be found has been 7-azidoactinomycin D. The usefulness of 7-azidoactinomycin D as a probe for examination of actinomycin D–DNA interactions relies on the ability of this analogue to mimic the parent compound in binding DNA prior to photolytic activation. In the absence of light, the equilibrium binding of 7-azidoactinomycin D is comparable in binding affinities, site exclusion size, and base sequence specificity to both the actinomycin D and 7-aminoactinomycin D parent molecules.[41]

Actinomycin D has been shown to bind double-stranded DNA via intercalation of the planar phenoxazone chromophore between adjacent base pairs of the DNA. This interaction has been shown to be highly sequence specific, with a marked preference for binding the d(GpC) step. The two cyclic pentapeptide side chains of actinomycin D are shown to reside in the minor groove of the DNA and thus are highly influenced by the DNA sequence of the bases that flank the intercalation site. These sequence specificities are retained by the photoreactive 7-azidoactinomycin D analogue. Complexation with DNA by actinomycin D and both 7-amino and 7-azido analogues results in comparable bathochromic and hypochromic shifts in the absorbance spectra of the drugs, as shown in Table 1. With calf thymus DNA and the synthetic alternating copolymers poly(dG–dC)·poly(dG–dC), a red shift of 30 nm is observed accompanied by a 30% decrease in absorbance. In contrast, no change in the visible spectra of any of the drugs is observed in the presence of the alternating copolymer poly(dA–dT)·poly(dA–dT), as shown in Figure 4.

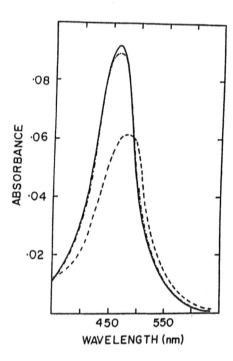

Figure 4. Spectral properties of 7-azidoactinomycin D binding to poly(dA–dT)·poly(dA–dT) (dashed line) and poly(dG–dC)·poly(dG–dC) (solid line).

VI. SEQUENCE SPECIFICITY OF ACTINOMYCIN D USING PHOTOAFFINITY LABELING

Photolysis of the 7-azidoactinomycin D–DNA complex results in photoactivation of the azide to the reactive nitrene and subsequent adduct formation. The ability to covalently bond the actinomycin D analogue provided us with a highly sensitive probe to examine sequence selectivity of actinomycin D. Covalent attachment of the drug allows sequence specificity studies to be performed at much lower drug concentrations than comparable studies using reversible binding compounds; thus, much greater specificity (i.e., high-affinity sites) can be observed. Second, as was reported by Helene and co-workers,[42] DNA photoadducts of the 3-azidoproflavine linked to DNA are alkali labile, thus circumventing the reliance on DNA cleavage agents such as DNase I or chemical cleavage, which are hampered by the need for single-hit kinetics and non-random DNA cleavage. Since the drug–DNA adduct is selectively cut at the site of covalent attachment of the drug by treatment of the adduct with the piperidine sequencing reaction, the resolution of the sequence specificity is markedly enhanced.

Results from these studies are consistent with a predominate preference for intercalation of the 7-azidoactinomycin D phenoxazone ring into the dGpdC step. The intercalation geometry is such that both the flanking G and C residues readily react to yield a piperidine-labile adduct. In addition, these studies revealed that the flanking sequences strongly influence the sequence specificities of the actinomycin D binding. Using this method, 7-azidoactinomycin D was observed to exhibit a strong binding preference for the d(TG$_n$T) motif, where n equals 2 to 4, with the strongest binding observed for the sequence d(TGGGT).[16] The sensitivity of this method, allowing observation of this sequence preference, had not been noted in previous studies.

VII. THE SHUFFLING HYPOTHESIS REVISITED

The mode through which actinomycin D exerts sequence specific binding to DNA was described by Fox and Waring in 1984 as initial interaction with non–sequence specific sites followed by linear diffusion of the drug along the DNA lattice and location of a preferred or high-affinity binding site. This hypothesis, known as the shuffling hypothesis, was based upon the time-dependent DNase I footprinting patterns observed for actinomycin D–DNA recognition.[23] However, the reversible nature of the actinomycin D–DNA interaction precluded a direct test of this hypothesis. The capacity for covalent bonding of actinomycin D to DNA provided a novel approach for a time-dependent analysis of the interaction of actinomycin D with DNA. In this study, the photoreactive actinomycin D analogue was allowed to interact with a restriction fragment for specified lengths of time ranging from 20 seconds to 45 minutes. Upon incubation, the drug–DNA complex was photolyzed, rendering the drug irreversibly bonded to the DNA. The modified DNA was then treated with hot piperidine and analyzed on a DNA sequencing gel. Results were quantitated by densitometric analysis. These studies reveal time-dependent movements from selected sites on the DNA lattice, with classically low affinity sites decreasing in reactivity (i.e., actinomycin D binding) and high-affinity sites increasing in reactivity. These studies provide strong support for the shuffling hypothesis by demonstrating that the locations of the actinomycin D on the DNA lattice change significantly with time.

VIII. CONCLUSIONS

These studies have shown photoaffinity labeling to be a powerful tool for examination of the interactions and mechanisms of the binding of ligands with nucleic acids. This approach allows reversible DNA binding agents to be converted to stable covalent adducts, permitting their precise location on the DNA lattice to be determined. In addition, insight into the mechanism(s) of ligand binding and sequence selectivity can also be obtained. Last, these photoaffinity labels can be used as probes to examine and characterize DNA structure and structural transitions. Earlier studies by Yielding and co-workers demonstrated the effectiveness of this method in correlating drug–DNA interactions with biological effects such as mutagenesis, carcinogenesis, and DNA repair processes.

REFERENCES

1. Waring, M. J. *Ann. Rev. Biochem.* **1981**, *50*, 159–192.
2. Chagas, C.; Pullman, B., Eds. *Molecular Mechanisms of Carcinogenic and Antitumor Activity,* Adenine Press: Schenectady, NY, 1987.
3. Knowles, J. R. *Acc. Chem. Res.* **1972**, *5*, 155–160.
4. Neilson, E.; Buchardt, O. *Photochem. Photobiol.* **1982**, *35*, 317–323.
5. Lwowski, W. *Ann. New York Acad. Sci.* **1980**, *346*, 491–500.
6. Graves, D. E.; Yielding, L. W.; Watkins, C. L.; Yielding, K. L. *Biochim. Biophys. Acta* **1974**, *479*, 98–104.

7. Hixon, S. C.; White, W. E., Jr.; Yielding, K. L. *Biochem. Biophys. Res. Commun.* **1975**, *66*, 31–35.

8. White, W .E., Jr.; Yielding, K. L. *Methods in Enzymology* **1977**, *46*, 644–659.

9. Fukunaga, M.; Yielding, K. L. *Mutation Research* **1981**, *80*, 91–97.

10. Hixon, S. C.; White, W. E., Jr.; Yielding, K. L. *J. Mol. Biol.* **1975**, *92*, 319–329.

11. Yielding, L. W.; White, W. E., Jr.; Yielding, K. L. *Mutation Research* **1976**, *34*, 351–358.

12. Yielding, L. W.; Brown, B. R.; Graves, D. E.; Yielding, K. L. *Mutation Research* **1979**, *63*, 225–232.

13. Yielding, L. W.; Graves, D. E.; Brown, B. R. *Biochem. Biophys. Res. Commun.* **1979**, *87*, 424–432.

14. Graves, D. E.; Watkins, C. L.; Yielding, L. W. *Biochemistry* **1981**, *20*, 1887–1893.

15. Turro, N. *Ann. New York Acad. Sci.* **1980**, *346*, 1–17.

16. Rill, R. L.; Marsch, G. A.; Graves, D. E. *J. Biomolec. Struct. Dynam.* **1989**, *7*, 591–605.

17. Gilbert, P. L.; Graves, D. E.; Chaires, J. B. *Biochemistry* **1991**, *30*, 10925–10931.

18. Gilbert, P. L.; Graves, D. E.; Britt, M.; Chaires, J. B. *Biochemistry* **1991**, *30*, 10931–10937.

19. Sobell, H. M.; Jain, S. C.; Sakore, T. D.; Nordman, C. E. *Nature New Biol.* **1971**, *231*, 200–205.

20. Sobell, H. M.; Jain, S. C. *J. Mol. Biol.* **1972**, *68*, 21–34.

21. Bailey, S.A. and Graves *Biochemistry* **1994**, *33*, 11493–11500.

22. Bailey, S.A. and Graves *Biochemistry* **1993**, *32*, 5881–5887.

23. Fox, K. R.; Waring, M. J. *Nucleic Acids Res.* **1986**, *14*, 2001–2014.

24. Fox, K. R.; Waring, M. J. *Nucleic Acids Res.* **1987**, *15*, 491–507.

25. Bailly, C.; Graves, D. E.; Ridge, G.; Waring, M. J. *Biochemistry* **1994**, *33*, 8736–8745.

26. Waring, M. J. *J. Mol. Biol.* **1965**, *13*, 269–282.

27. Waring, M. J. *J. Mol. Biol.* **1972**, *54*, 247–279.

28. Cantrell, C. E.; Yielding, K. L. *Photochem. Photobiol.* **1980**, *32*, 613–619.

29. Cox, B. A.; Firth, W. J.; Hickman, S.; Klotz, F. B.; Yielding, L. W.; Yielding, K. L. *J. Parasitol.* **1981**, *67*, 410–416.

30. Garland, F., Graves, D. E., Yielding, L. W., Cheung, H. C. *Biochemistry* **1980**, *19*, 3221–3226.

31. Bolton, P. H.; Kearns, D. R. *Nucleic Acids Res.* **1978**, *5*, 4891–4903.

32. Kulkarni, M. S.; Yielding, K. L. *Chem.-Biol. Interact.* **1985**, *56*, 89–99.

33. Coffman, G. L.; Gaubatz, J. W.; Yielding, K. L.; Yielding, L. W. *J. Biol. Chem.* **1982**, *257*, 13205–13207.

34. Lane, M. J.; Dabrowiak, J. C.; Vournakis, J. N. (1983) *Proc. Natl. Acad. Sci. U.S.A.* **1983**, *80*, 3260–3264.

35. Goodisman, J.; Dabrowiak, J. C. In *Advances in DNA Sequence Specific Agents*; Hurley, L. H., Ed.; JAI Press: Greenwich, CT, 1992; pp. 25–50.

36. Rill, R. L.; Marsch, G. A.; Graves, D. E. *Nucleic Acids Res.* **1995** *23*, 1252–1259.

37. Reinhardt, C. G.; Krugh, T. R. *Biochemistry* **1978**, *17*, 4845–4854.

38. Kastrup, R. V.; Young, M. A.; Krugh, T. R. *Biochemistry* **1978**, *17*, 4855–4865.
39. Pohl, F. M.; Jovin, T. M. *J. Mol. Biol.* **1972**, *67*, 375–396.
40. Pohl, F. M. *Cold Spring Harbor Symp. Quant. Biol.* **1983**, *47*, 113–118.
41. Graves, D. E.; Wadkins, R. M. *J. Biol. Chem.* **1989**, *264*, 7262–7266.
42. Le Doan, T.; Perrouault, L.; Praseuth, D.; Habhoub, N.; Decout, J. L.; Thuong, N. T.; Lhomme, J.; Helene, C. *Nucleic Acids Res.* **1987**, *15*, 7749–7760.

DNA BINDING OF DINUCLEAR
PLATINUM COMPLEXES

Nicholas Farrell

Advances in DNA Sequence Specific Agents
Volume 2, pages 187–216.
Copyright © 1996 by JAI Press Inc.
All rights of reproduction in any form reserved.
ISBN: 1-55938-166-3

I. INTRODUCTION

The predominant role of DNA in replication and transcription makes it a principal target of natural and synthetic cytotoxins. The chemistry and biochemistry of DNA related to its cellular function suggest a rich variety of approaches for inhibition of DNA synthesis. The sequence specificity, topological constraints, and localized conformational demands necessary for successful gene regulation are all properties capable of modification by binding of small molecules and drugs to DNA. The structural diversity of clinically used anticancer agents[1] and their discrete modes of action is testament to the variety of pathways available for the selective inhibition of DNA synthesis.

This chapter summarizes the design and development of a new class of DNA-binding antitumor agents based on a dinuclear platinum structure, bis(platinum) complexes. The advent of the simple coordination complexes cisplatin [cis-PtCl$_2$(NH$_3$)$_2$], cis-DDP, and carboplatin [Pt(CBDCA)(NH$_3$)$_2$] into cancer chemotherapy regimens has made a major impact in cancer treatment. Early studies showed that the $trans$-DDP isomer was devoid of antitumor activity.[2] The avid binding of both isomers to DNA resulted in considerable emphasis on the understanding of the specific modes of DNA binding of the active cis-isomer, and its mechanism of action is now known in some detail.[3,4] The toxic lesions resulting from the cis-DDP–DNA interaction are considered to be bifunctional, with the 1,2-intrastrand d(GpG) adduct predominating together with the occurrence of lesser proportions of d(ApG) and 1,3-d(GpNpG) intrastrand cross-links as well as interstrand GG cross-links (Figure 1). How the conformational changes induced by cis-DDP and its congeners affect gene-specific damage and repair, as well as the nature of protein recognition of cis-DDP-damaged DNA, are two of many aspects that remain to be fully elucidated.[5,6]

The empirical structure–activity relationships delineated for platinum complexes stressed the necessity for neutral complexes of the cis-[PtX$_2$(amine)$_2$] structure, where X is a leaving group, such as chloride, and amine represents ammonia or a primary monodentate or bidentate amine. Analogues based on the cis-

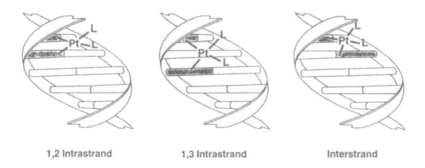

| 1,2 Intrastrand | 1,3 Intrastrand | Interstrand |

Figure 1. Schematic limiting modes for *cis*-DDP–DNA adducts. The 1,2-intrastrand adducts are formed between d(GpG) and d(ApG) base pairs, the 1,3-intrastrand is formed in a d(GpNpG) sequence, and the interstrand cross-link is formed between guanine atoms on adjacent d(GC) base pairs.

DDP structure have not, however, shown a greatly altered spectrum of clinical efficacy in comparison to the parent drug.[7,8] The mechanistic explanation for this finding is that all *cis*-[PtX$_2$(amine)$_2$] compounds produce an array of adducts very similar to those of *cis*-DDP. Thus, the biological consequences are also expected to be similar.

As a synthetic agent, it is of course unlikely that the structure of *cis*-DDP, and the subsequent bifunctional DNA binding, has been optimized for its medical application. The cytotoxic effects of *cis*-DDP are probably best considered as resulting from the cumulative effects of the various structurally distinct adducts. Our interest in determining if these adducts were uniquely effective within platinum chemistry led to the study of dinuclear complexes, a large class of general formula [{PtCl$_m$(NH$_3$)$_{3-m}$}$_2$ (H$_2$N–R–NH$_2$)]$^{2(2-m)+}$ (*m* = 0–3 and R is a linear or substituted aliphatic linker). Their biological properties are highlighted by high antitumor activity in both *cis*-DDP-sensitive and resistant cells[9,10] and unique modes of DNA binding inaccessible to the mononuclear complexes.[11,12] Their study represents the first clearly demonstrated alternative mechanism of action for platinum complexes.

II. DNA INTERACTIONS OF DINUCLEAR PLATINUM COMPLEXES

Figure 2 shows the structures of the principal dinuclear or bis(platinum) complexes under investigation in our laboratories.

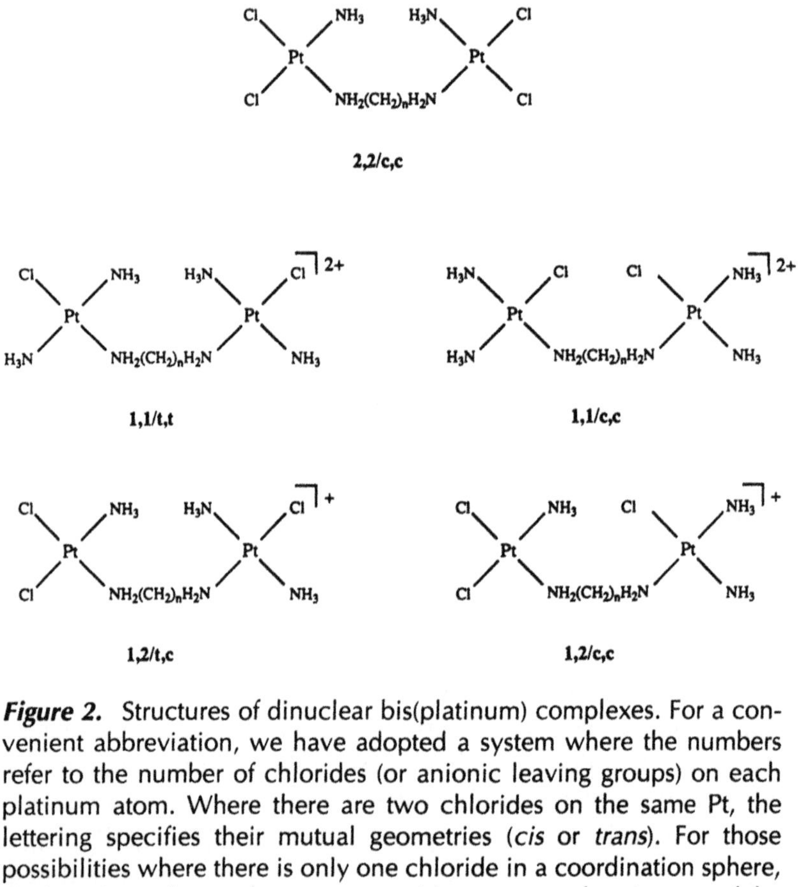

2,2/c,c

1,1/t,t   **1,1/c,c**

1,2/t,c   **1,2/c,c**

Figure 2. Structures of dinuclear bis(platinum) complexes. For a convenient abbreviation, we have adopted a system where the numbers refer to the number of chlorides (or anionic leaving groups) on each platinum atom. Where there are two chlorides on the same Pt, the lettering specifies their mutual geometries (*cis* or *trans*). For those possibilities where there is only one chloride in a coordination sphere, the lettering refers to the geometry with respect to the nitrogen of the bridging diamine. Once these two parameters are specified, the geometry of the overall complex is automatically fixed. Thus [{cis-PtCl$_2$(NH$_3$)$_2$}$_2$H$_2$N(CH$_2$)$_n$NH$_2$] is 2,2/c,c, [{trans-PtCl(NH$_3$)$_2$}$_2$H$_2$N-(CH$_2$)$_n$NH$_2$]Cl$_2$ is 1,1/t,t, etc. The 1,2/t,c denomination therefore refers to a *cis*-DDP unit and a monofunctional unit with Cl *trans* to the diamine bridge.

The basic dinuclear structure allows for considerable flexibility in the design of specific DNA–DNA and DNA–protein binding agents. In principle, the sources of specificity include chain length and steric effects of the diamine backbone, the nature of the other ligands attached to platinum, the coordination sphere (monofunctional and/or bifunctional), and coordination geometry (whether the leaving groups are *cis* or *trans* to the diamine bridge). Figure 3 shows in a schematic manner the limiting DNA binding modes available to these species. Upon initial monofunctional binding, further reaction of the second Pt site may lead to a series of structurally distinct adducts, including

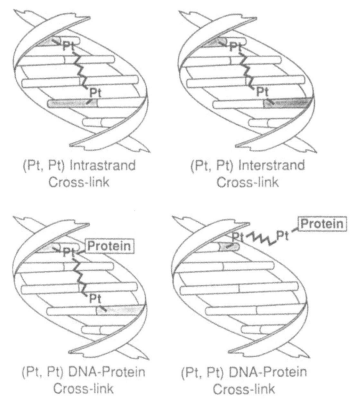

(Pt, Pt) Intrastrand
Cross-link

(Pt, Pt) Interstrand
Cross-link

(Pt, Pt) DNA-Protein
Cross-link

(Pt, Pt) DNA-Protein
Cross-link

Figure 3. Schematic limiting modes for bis(platinum) DNA–DNA and DNA–protein cross-links.

DNA–(Pt,Pt) interstrand cross-links by binding of one Pt atom to each strand of DNA, (Pt,Pt) intrastrand cross-links by binding of the two Pt atoms to the same strand, and DNA–protein cross-links. The DNA–DNA and DNA–protein binding modes shown in Figure 3 are minimal in the sense that they require only one leaving group (chloride) on each Pt atom, producing bifunctional adducts on DNA. A trifunctional or tetrafunctional bis(platinum) complex, containing cis-DDP units, will be capable of further interaction with DNA. A further modification to enhance selectivity is to utilize the individual kinetic and thermodynamic properties of two different metals incorporated into a heterodinuclear structure. The properties of a model heterodinuclear complex containing both Pt and Ru as a DNA–DNA and DNA–protein cross-linking agent have been reported.[13,14] In this review, we will first outline the DNA-binding properties of bifunctional dinuclear complexes and contrast their properties with the binding of cis- and trans-DDP. Then the features of trifunctional and tetrafunctional complexes will be briefly summarized.

III. BIFUNCTIONAL DINUCLEAR PLATINUM COMPLEXES

Dinuclear complexes capable of bifunctional binding are of general formula $[\{PtCl(NH_3)_2\}_2(H_2N-R-NH_2)]^{2+}$. The chloride leaving groups are either cis or trans to the diamine bridge, giving 1,1/c,c and 1,1/t,t geometric isomers, respectively (see Figure 2). The 1,1/c,c case is sterically more demanding with respect to substitution of the chloride leaving groups, and this situation is reflected in differences in DNA binding and biological activity of the two isomers, although the overall profiles are similar (Table 1). Note that the potent antitumor activity of these species[15] violates all known structure–activity relationships: the complexes are charged and contain formally monofunctional platinum coordination spheres. It is instructive to compare their DNA-binding properties with cis- and trans-DDP.

Table 1. Selected In Vivo Activity of Charged Bifunctional Bis(platinum) Complexes in Murine Leukemia and Human Tumor Xenografts[a]

	L1210/DDP (iv,iv d1,4,7)[b]	L1210/DDP (ip,ip d1,5,9)	IGROV1 (ip,ip d3,7,11)	A2780/DDP (sc/iv q4d × 3)
		dose, %T/C[c]		dose, TWI%[c]
cis-DDP	3, 95	4, 103	3, 226	3, 38%
1,1/t,t (n = 6)	2, 143	2, 194(2/8)[d]	1, 217	2, 77%
1,1/c,c (n = 4)	—	1.5, 215	—	2, 72%
1,1/c,c (n = 6)	1.5, 94	2, 122	1.5, 230	2, 54%

Notes: [a]L1210/DDP is murine leukemia resistant to *cis*-DDP; IGROV1 is human ovarian carcinoma sensitive to *cis*-DDP; A2780/DPP is human ovarian carcinoma resistant to *cis*-DDP.

[b]Tumor transplant route/drug administration, dose schedule.

[c]Results are presented as dose (mg/kg), %T/C for L1210/DDP and IGROV1 and dose (mg/kg), %TWI for A2780/DDP.

[d]Thirty-day survivors in parentheses. See Figure 2 for structures. Data taken from Reference 15. The results show that dinuclear platinum complexes display high antitumor activity in vivo in cell lines both sensitive and resistant to *cis*-DDP.

A. Sequence Specificity of Adduct Formation; Kinetics of Binding of Dinuclear Platinum Complexes

By use of a $5'$-^{32}P-labeled 49-bp duplex, the $3'$ exonuclease activity of T4-DNA polymerase, and DNA sequencing gels, alternating purine–pyrimidine GCGC sequences were identified as binding sites for [{*trans*-PtCl(NH$_3$)$_2$}$_2$H$_2$N(CH$_2$)$_n$NH$_2$]$^{2+}$ complexes.[16] These binding sites were seen in addition to the GG binding sites observed for both the dinuclear compound and *cis*-DDP. The preferred binding sites within the sequence below are indicated by the bases accentuated:

$5'$-GACTACTTG$_9$G$_{10}$TACACTGAC$_{19}$GCG$_{22}$A-
GCTC$_{27}$GCGG$_{31}$AAGCTCATTCCAGTGCGC-$3'$

We have used this sequence for many of the analyses presented in this summary because it presents a combination of *cis*-DDP (G$_9$G$_{10}$, G$_{30}$G$_{31}$) and alternative (C$_{19}$–G$_{22}$ and G$_{28}$–G$_{31}$) binding sites. Early studies indicated that bis(platinum) complexes reacted faster than *cis*-DDP with DNA.[17] The reactions of bis(platinum) complexes with the self-complementary 5'-ATA-TAT**GCGC**ATATAT-3' (GCGC) and 5'-ATATAT**GGCC**ATATAT-3' (GGCC) were examined to determine the kinetic preferences between the two defined sequences.[18] The reaction of 1 μM of the oligonucleotides with 0.1–5.0-μM concentrations of [{*trans*-PtCl(NH$_3$)$_2$}$_2$H$_2$N(CH$_2$)$_4$NH$_2$]Cl$_2$ for 1 h gave two principal bands, attributed to binding of the two independent guanines in each sequence (Figure 4). This assignment is reasonable given the demonstrated kinetic preference for guanine N-7 binding in Pt–NH$_3$ complexes.[19] The rate of disappearance of free oligonucleotide along with the relative concentrations of the two products allowed calculation of an initial rate of reaction. The rate of formation of the two products is not equal, suggesting that reactivity of the guanines is dependent on other factors besides nucleophilicity. The two guanine sites in each oligonucleotide are flanked by different bases: 5'-T<u>G</u>G-3' and 5'-G<u>G</u>C-3' for GGCC and 5'-T<u>G</u>C-3' and 5'-C<u>G</u>C-3' for GCGC. The different chemical environments around the individual N-7 sites are likely to affect the binding preferences. The GGCC oligonucleotide reacted faster than the GCGC counterpart (Figure 4).

Because GCGC is a non–cisplatin-like site, it was of interest to compare the preferences of mononuclear and dinuclear complexes within this sequence. The rates of reaction were 1,1/t,t > [Pt(dien)]$^{2+}$ >> *cis*-DDP. The second-order rate constants for reaction with [{*trans*-PtCl(NH$_3$)$_2$}$_2$H$_2$N(CH$_2$)$_4$NH$_2$]$^{2+}$ were 60.0 and 6.1 M^{-1}s^{-1} for the individual guanines, in contrast to values of 21.0 and 6.1 M^{-1}s^{-1} for [PtCl(dien)]$^+$. Because only one substitution reaction (monofunctional binding) occurs on each platinum center in both complexes, the increased charge of the dinuclear derivative (2+ vs. 1+) may contribute to the enhanced initial reactivity. Electrostatic effects may play an important role in initial drug–DNA interactions, and, in agreement with this point, increasing concentrations of NaClO$_4$ inhibited the reaction of the dinuclear complexes with fewer products visualized under high-salt conditions. For both sequences, the dinuclear complex was at least an order of magnitude more reactive than *cis*-DDP.[18]

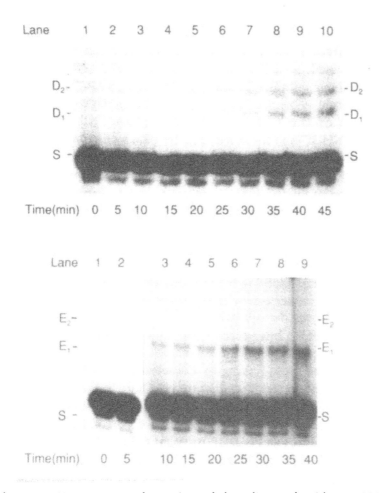

Figure 4. Time course of reaction of the oligonucleotides 5'-ATA-TA**GCGC**ATATAT-3' (Panel D) and 5'-ATATA**GGCC**ATATAT-3' (Panel E) with [{*trans*-PtCl(NH$_3$)$_2$}$_2$H$_2$N(CH$_2$)$_4$NH$_2$]$^{2+}$ on denaturing polyacrylamide gels. The oligonucleotides were reacted at 0.1 µM with 0.2 µM of complex. S refers to starting oligonucleotide and D$_1$, D$_2$, E$_1$, and E$_2$ refer to the appearance of bands corresponding to platinated guanines within the central sequences of GCGC and GGCC, respectively. The rate of reaction can be calculated by measurement of radioactive counts in each band and total radioactivity. See Reference 18 for experimental details.

The kinetic results indicate that differences in antitumor activity between dinuclear and mononuclear platinum complexes are dictated, to a first degree, by the nature and *structure* of the adducts within a similar sequence (i.e., GGCC) rather than a different sequence specificity. Further analysis of the time course of the reaction between [{*trans*-PtCl(NH$_3$)$_2$}$_2$H$_2$N-(CH$_2$)$_4$NH$_2$]$^{2+}$ and the oligomers, by denaturing PAGE, indicate that an array of bis(Pt) adducts are eventually formed, including several structurally distinct (Pt,Pt) interstrand cross-links. The presence of GCGC runs in many promoter sequences of DNA suggests that dinuclear species may be designed to specifically modify such regions. Indeed, the use of planar ligands such as pyridine in place of NH$_3$ in complexes such as *trans*-[PtCl$_2$(py)$_2$] does confer GC specificity to DNA binding.[20] Incorporation of this structural motif into the dinuclear structure should allow for further enhancement of sequence specificity.

B. The (Pt,Pt) Interstrand Cross-Link; Sequence Assignment

Interstrand cross-linking by dinuclear platinum complexes is significantly greater than that by *cis*-DDP. In the 1,1/t,t series, interstrand cross-linking falls off dramatically in the order $n = 4 > 3 \gg 2$. Interestingly, this is the order of cytotoxicity of these complexes in *cis*-DDP resistant cells.[21] The cross-linking is also dependent on the geometry around the Pt atom. *cis*-DDP, and indeed most alkylating agents, gives rise to only 1,2-interstrand cross-links. In contrast, both the length and the flexibility of the diamine chain in dinuclear compounds allow the targeting of much larger DNA sequences for cross-link formation.[22,23] For binding between guanines on opposite strands, in addition to 1,2 cross-links, 1,3 and 1,4 interstrand cross-links are also possible. In 1,3 and 1,4 cross-links, the guanines are separated by one and two base pairs, respectively, whereas a 1,2 cross-link is formed from guanines on neighboring base pairs. The detailed mechanistic understanding of bis(platinum) complexes requires knowledge of the sequence specificity and structure of the interstrand cross-links.

Sequence specificity of interstrand cross-link formation has been assayed by a number of methods.[24–26] However, most DNA-targeting drugs and agents produce more than one type of DNA adduct, making it difficult to study the effects of a specific lesion on random sequence DNA. This is especially true in the case of mononuclear platinum complexes where interstrand cross-links present only a minor portion of the total adducts. For *cis*-DDP, the selective reaction of the various types of Pt–DNA adducts with NaCN may be exploited to obtain information on interstrand cross-links.[27,28] The 1,2-d(GC) cross-link has been considered most favored over the alternative d(CG) structure.[28]

The sequence assignment of intrastrand and interstrand cross-links caused by platinum complexes has been aided by development of a new assay, taking advantage of the fact that 3'–5' exonuclease digestion of randomly platinated DNA produces a pool of fragments of different length.[29] The practical features of the assay are outlined in Figure 5. The inhibition of 3'–5' exonuclease activity of enzymes by DNA adducts may be used to monitor the binding sites of small molecules. The exonuclease activity is usually arrested 1–3 base pairs away from the adduct, and enzymatic digestion of the damaged DNA produces a pool of fragments of varying lengths, owing to the inhibition by the array of adducts (Figure 5b and c).[30–32] This treatment allows identification of the binding site but not the nature of the adducts impeding the exonuclease scission. Information on the structural nature of the adducts may be obtained in the following manner. After an appropriate period of digestion, the samples are immediately heated to eliminate the exonuclease activity. Upon cooling to 37 °C, DNA polymerase and an excess of dNTPs are added to allow any primed fragments to be extended (Figure 5d). The DNA fragments containing only monofunctional and intrastrand adducts are most likely single-stranded at this stage. In contrast, those fragments with interstrand cross-links that remain complementary in the proximity of lesions (Figure 5c) can be extended by the action of a DNA polymerase (Figure 5d). As a result, this extension increases the size of the oligonucleotide fragments (Figure 5e), which may be evidenced by a more slowly migrating band on a se-

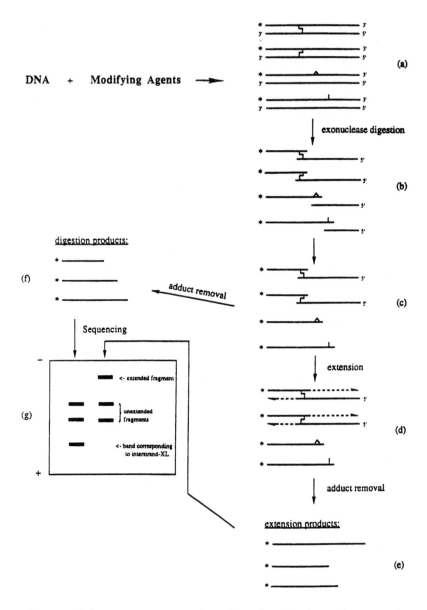

Figure 5. Schematic representation of the digestion/extension assay for direct assignment of sequence specificity of interstrand cross-linking. See text and Reference 29 for details.

quence gel (Figure 5g). The band corresponding to the digested cross-link decreases in intensity. Therefore, comparison of a sequencing gel after digestion only and after the digestion/extension treatment should show the disappearance, or diminished intensity, of those interstrand cross-links with complementary ends. No changes should be apparent for fragments not due to cross-links, because these are not extendable.

A demonstration of the utility of this approach can be given by a summary of results from both *cis*-DDP and the dinuclear complexes. In the $G_{28}C_{29}G_{30}G_{31}$ sequence of the 49-bp oligomer described previously, the most likely binding site for *cis*-DDP would be a $(G_{30}G_{31})$ intrastrand cross-link. A secondary possibility would be the $(G_{28}C_{29}G_{30})$ intrastrand cross-link. However, examination of the gel patterns before and after the digestion/extension treatment described in Figure 5 showed, surprisingly, the presence of an interstrand cross-link rather than the possible 1,2-d(GG) or 1,3-d(GCG) adducts. In the case of the $d(G_{28}C_{29}G_{30}G_{31})$, the interstrand cross-link may form from either G_{28} or G_{30} [one of the guanosine nucleosides of the possible d(GG) *intrastrand* adduct] to the G'_{29} of the opposite strand. The susceptibility of GG sites to binding by *cis*-DDP is subject to local sequence and conformational effects.[33,34] Even where GG binding is indicated (as judged by assays such as exonuclease inhibition), the results from the digestion/extension assay show that assignments of stop sites to intrastrand cross-links should be made with caution. The sequencing results raise the possibility of interstrand cross-link formation in "GC boxes" such as that of d(GGGCGG), a regulatory sequence in tumor virus SV40 DNA and a very specific target of *cis*-DDP.[35,36]

For regions of interstrand cross-linking in bis(platinum) complexes, the increased frequency of such cross-links is readily identified, and the ability of a dinuclear bis(platinum) complex to target longer sequences of DNA was confirmed. The detailed description of their structure is complicated by the possibility of multiple interstrand cross-links being formed from one Pt binding site on one strand to more than one site on the other. The analysis is complicated further by the possibility that interstrand cross-linking may occur both to the 5' and 3' sides

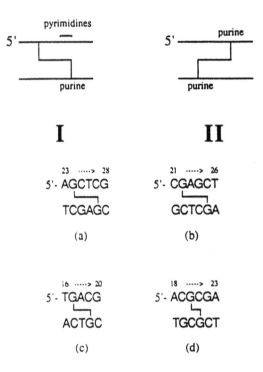

Figure 6. Possible DNA–DNA interstrand cross-links formed by [{trans-PtCl(NH$_3$)$_2$}$_2$H$_2$N(CH$_2$)$_4$NH$_2$]Cl$_2$. Type I is oriented in the preferred 5' → 5' direction whereas Type II is the less likely 3' → 3' direction. Sequences (a)–(d) represent sites of (1,2), (1,3), and (1,4) interstrand cross-links within the 49-bp polymer 5'-GACTACTTGG$_{10}$TA-CACTGACG$_{20}$ CGAGCTCGCG$_{30}$ GAAGCTCATT$_{40}$ CCAGTGCGC-3', where 5' and 3' stand for the phosphate and hydroxyl ends of the double-stranded DNA, respectively.

of the Pt binding site (types I and II, Figure 6). The 5'–5' (type I) cross-links are most likely favored over other possibilities, considering that a 5'-dG reacts faster with *cis*-DDP than does the 3'-dG base in short d(G$_n$) sequences.[37,38] Although specific sequences are more difficult to assign, an array of 1,2, 1,3, and 1,4 cross-links are identifiable, by consideration of the guanine binding sites (the most favored of the four bases) on both strands of a DNA duplex (Figure 6). The nature of the intervening bases, with the possibility of cross-linking alternating purine–pyrimidine sequences, will have consequences for the global conformational changes resulting from the final adduct. The formation of long-range cross-links is unusual in the major groove of DNA, and the orientation of the dinuclear chain could present a very efficient block to protein recognition within the major groove.

C. The (Pt,Pt) Intrastrand Cross-Link; Model NMR Studies

In a dinuclear complex, initial monofunctional binding of one center leads to possible (Pt,Pt) interstrand cross-linking by binding of the second center to the opposite strand or, alternatively, (Pt,Pt) intrastrand cross-linking by binding of the second Pt atom to the same strand (see Figure 3). These preferences may be controlled by the geometry of the bis(platinum) complex.

The sequencing assay described in Figure 5 for [{*trans*-PtCl(NH$_3$)$_2$}$_2$H$_2$N(CH$_2$)$_4$NH$_2$]$^{2+}$ implied (by the presence of non-extendable stop sites for digestion corresponding to these sequences) the formation of (Pt,Pt) intrastrand cross-links at d(G$_9$G$_{10}$) and d(G$_{30}$G$_{31}$). This adduct thus represents the direct structural analogue of the major adduct of *cis*-DDP. Within the 49-bp sequence we have used for these studies, the d(G$_9$G$_{10}$) is diagnostic for the intrastrand cross-link as it is flanked by nonbinding T bases, and assignment to G binding is straightforward. In contrast, the d(G$_{30}$G$_{31}$) site is part of a possible interstrand cross-linking sequence, as discussed previously. An interesting feature was observed when the sequence specificity of [{*cis*-PtCl(NH$_3$)$_2$}$_2$H$_2$N(CH$_2$)$_4$NH$_2$]$^{2+}$ was compared with that of its *trans*-isomer.[39] No d(G$_9$G$_{10}$) stop site was observed, and this band is unique to the 1,1/t,t complex (compare lanes 3 and 4 in Figure 7). Other binding sites, especially those attributable to interstrand cross-linking regions, are observed for the 1,1/c,c complex.

The through-bond Cl–Cl distances are approximately 9 and 14 Å for the $n = 4$ 1,1/c,c and 1,1/t,t isomers, respectively. These distances compare well at first glance with the expected 4 Å between adjacent guanine N-7 atoms in a B-form GGCC sequence. However, only the diamine chain in the 1,1/t,t case

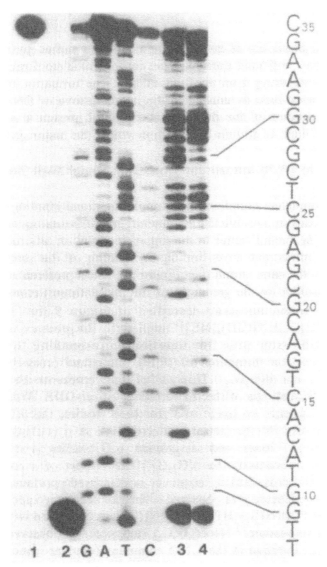

Figure 7. Sequencing gel of the 49-bp 5'- GACTACTTGG$_{10}$TA-CACTGACG$_{20}$ CGAGCTCGCG$_{30}$ GAAGCTCATT$_{40}$ CCAGTGCGC-3' oligonucleotide damaged by dinuclear platinum complexes. Lane 1 is unmodified DNA; lane 2 is DNA treated with *cis*-DDP, and G, A, T, and C represent the products of Maxam–Gilbert sequencing for the bases indicated; lane 3 is DNA modified by [{*trans*-PtCl(NH$_3$)$_2$}$_2$-H$_2$N(CH$_2$)$_4$NH$_2$]$^{2+}$; and lane 4 is the same modified by its geometric isomer [{*cis*-PtCl(NH$_3$)$_2$}$_2$H$_2$N(CH$_2$)$_4$NH$_2$]$^{2+}$, as in Figure 3.

can be rotated around the two Pt square planes to allow simultaneous Pt binding at both guanine N-7 positions. The greater steric demands of two Cl leaving groups *cis* to the diamine bridge prevent a favorable orientation of the linker chain to permit simultaneous binding at both guanines, at least for the 1,4-butanediamine linker. Thus, geometric isomerism controls the relative formation of (Pt,Pt) interstrand vs. (Pt,Pt) intrastrand cross-link formation.

Model studies with the dinucleotide d(GpG) confirmed the formation of (Pt,Pt) intrastrand cross-links with the dinuclear $[\{trans\text{-}PtCl(NH_3)_2\}_2H_2N(CH_2)_6NH_2]^{2+}$.[40] We are currently investigating the detailed structure. The important spectral features derived from study of the NMR spectrum to date are the separation of the two independent (5′ and 3′) H-8 protons and the sugar conformation. A COSY analysis indicates that the 5′ sugar becomes more N-type. In contrast, both NMR and X-ray crystallography show that the 5′G sugar in the *cis*-DDP chelate adopts an exclusively N-type pucker.[41,42] The H-8 separation is also dependent on chain length, implying that the chain length affects the relative orientation of the guanine bases.[43] Kinetic studies showed that the rate of formation (and proportion relative to other products) of the chelate is dependent on the chain length with $n = 6 > 4 > 2$. Molecular modeling showed that the (Pt,Pt) intrastrand cross-link results in kinking of the backbone. The N-type pucker observed for *cis*-DDP and *cis*-DDP analogues[44] is due to the steric constraints of bifunctional binding from one Pt center. Monofunctional binding to oligonucleotide sequences from one Pt center as from $[PtCl(dien)]^+$ does not require any major changes in sugar pucker because the flexibility of the sugar–phosphate backbone is not restricted.[45] The (Pt,Pt) intrastrand cross-link appears to produce a less rigid adduct than is formed with *cis*-DDP. This is a result not only of the fact that one monofunctional Pt center binds to each of the two guanines but also of the flexibility of the diamine chain.

The conformational differences between the (Pt,Pt) intrastrand adduct and the intrastrand d(GpG) cross-link of *cis*-DDP could have important biological consequences. The latter is not removed as efficiently as interstrand cross-links in gene-specific repair,[46] and other workers have confirmed that site-defined d(GpG)–cis-DDP adducts are refractory to repair in cell extracts.[47] Differential repair of the (Pt,Pt) intrastrand adduct relative to that of cis-DDP could affect, for example, cytotoxicity in cis-DDP resistant cells where enhanced DNA repair is implicated in the mechanism of resistance.[48]

D. DNA Conformational Changes Induced by Dinuclear Platinum Complexes

Bis(platinum) complexes containing monofunctional coordination spheres are potent antitumor agents, whereas their mononuclear analogues such as $[PtCl(NH_3)_3]^+$ and $[PtCl(dien)]^+$ are inactive. The mechanistic reason for the latter observation is that monofunctional binding, as monitored by NMR[49] and physical and biochemical studies,[50] causes only localized conformational changes without any kinking of the helix. In agreement with this, monofunctional binding is not an efficient inhibitor of polymerase excision.[32] The dinuclear complexes achieve bifunctional binding from two platinum centers. If bifunctional complexes produce long-range cross-links, it might be supposed that protein processing of this damage would proceed as for the monofunctional mononuclear complex. However, lesions caused by the dinuclear complexes are more effective at inhibition of replication and transcription and are not repaired in the same manner as the purely monofunctional binding of $[PtCl(NH_3)_3]^+$.[51] These considerations, plus the dramatic antitumor activity shown in Table 1, clearly show that the effect of bifunctional binding of the dinuclear complexes is not simply the independent sum of two monofunctional units, as has been suggested.[52] The previous sections have also demonstrated the different adduct profile of these agents in comparison to that of cis-DDP. What then is the nature of the global conformational changes caused by (Pt,Pt) interstrand and/or (Pt,Pt) intrastrand adducts?

1. B–Z DNA Conformational Changes

One interesting demonstration of the differences between mononuclear and dinuclear complexes involves stabilization of Z-form DNA. The dinuclear complexes [{*cis/trans*-PtCl(NH$_3$)$_2$}$_2$H$_2$N(CH$_2$)$_n$NH$_2$]$^{2+}$ are especially efficient at inducing the B–Z transition in poly(dG–dC)·poly(dG–dC).[53] We have found that as few as one 1,1/t,t bis(Pt) lesion per 25 bases is needed for complete conversion to the Z form. Little change is seen in the conformation of poly(dG)·poly(dC) at equivalent r_b. The effect of different adduct structures on conformational changes within a similar sequence may be appreciated by the fact that *cis*-DDP stabilizes B-form poly(dG–dC)·poly(dG–dC).[54,55] The monofunctional [PtCl(dien)]$^+$ also facilitates the B–Z DNA transition at low r_b but does not by itself fully induce the conformational change.[55] This is in accord with the observations that metal–amine complexes with *facial* arrangement of NH$_3$ groups, such as [Co(NH$_3$)$_6$]$^{3+}$ and [Ru(NH$_3$)$_6$]$^{3+}$, are efficient at stabilizing Z-form DNA.[56,57] The mere arrangement of facial NH$_3$ groups is not, by itself, sufficient, and charge and size are also important factors. Small square-planar complexes such as [Pt(NH$_3$)$_4$]$^{2+}$ do not induce the B–Z conversion under any conditions and tend simply to precipitate the DNA with increased concentration.[58] As indicated by the crystal structures of the Co and Ru salts with hexanucleotides, the octahedral arrangement is probably optimal for the array of H-bonding contacts necessary for the B–Z conversion.

To examine the effect of charge and structure on this transformation, we studied the tetraamine complexes[59] {[Pt(NH$_3$)$_3$]$_2$H$_2$N(CH$_2$)$_n$NH$_2$}$^{4+}$ and their mononuclear analogue [Pt(NH$_3$)$_4$]$^{2+}$, neither of which are capable of covalent binding (Scheme 1). On a molar basis, the 4+ species were more effective than [Co(NH$_3$)$_6$]$^{3+}$ in causing the B–Z transition.[60,61] Further studies with a series of dinuclear compounds have delineated the mechanistic scheme for Z-DNA formation shown in Scheme 2.[18,53] In this scheme, Z$_i$, Z$_{ii}$, and Z$_{iii}$ refer to different forms of left-handed DNA as indicated by circular dichroism spectroscopy.[18,53] Thus, Z$_i$ is induced by purely electrostatic in-

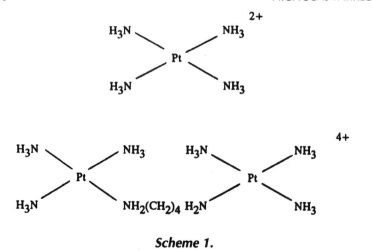

Scheme 1.

teractions from tetraamine bis(platinum) cations, Z_{ii} is induced by monofunctional binding from only one Pt of a dinuclear complex, and Z_{iii} is due to bifunctional binding and cross-linking.

Bifunctional binding is not a prerequisite for the B–Z transition. Rather, bifunctional binding or interstrand cross-linking "locks" the DNA in the left-handed conformation. In agreement with this, ethidium bromide *does not* reverse the Z-conformation caused by the interstrand cross-linking agents.[62] The mononuclear $[Pt(NH_3)_4]^{2+}$ cation does not, as stated, cause the B–Z change, even at 10 times the concentration of the 4+ species. The two 2+ units in the dinuclear cation must then act in some

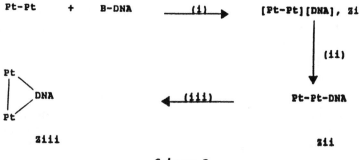

Scheme 2.

cooperative fashion to achieve the Z-form induction, possibly aided by a contribution to hydrophobic interactions from the diamine chain. In this instance, there is a structural similarity to the polyamines, whose Z-induction has been well studied.[63] Many agents are capable of causing the B–Z transition but this is mostly reversible. The cross-linking to Z-form DNA (induced by heating in the presence of Mn^{2+}) of DL-diepoxybutane inhibits the interconversion between B and Z forms.[64] Interstrand cross-linking that induces the B–Z conformation would be expected, therefore, to be difficult to reverse. The ability to lock the Z-DNA conformation could have important consequences in regulation of transcription, where the relevance of Z-DNA regions has been implicated.[65,66]

2. Unwinding and Bending of DNA by Dinuclear Platinum Complexes

The result of drug–DNA binding is manifested in effects on replication and transcription. Conformational changes may affect protein recognition necessary for a host of DNA-related processes. In this respect, repair proteins such as those involved in the UvrABC complex appear to recognize bulky adducts, independent of the chemical origin (alkylating agent, platinum complex, etc.) of the lesion.[67]

The global conformational changes of importance to protein recognition of DNA are unwinding and bending. At equivalent r_b, *trans*-DDP is significantly less effective at unwinding than is the *cis*-isomer.[68,69] For platinum complexes, the helix unwinding, as measured by topological unwinding of supercoiled plasmids, is dependent on the geometry of the complex and the structure of the specific adduct.[52] A species such as $[PtCl(dien)]^+$ does not unwind supercoiled DNA to any great extent, implying that the localized disruption as noted previously is not transmitted over a wide number of base pairs. The unwinding induced by the pair of $[\{PtCl(NH_3)\}_2H_2N(CH_2)_4NH_2]^{2+}$ isomers is essentially identical. The unwinding angles are similar to that found for *cis*-DDP and are typical of bifunctional binding (see Table 2). It can be seen that unwinding equivalent to that of

Table 2. Comparison of Effects of Bifunctional DNA Binding of
Mononuclear and Dinuclear Platinum Complexes

Complex	Unwinding Angle[a]	Bending Angle[b]	SSRP (HMG) Recognition[c]	Interstrand Cross-Linking	Intrastrand Cross-Linking	B–Z Induction
cis-DDP	11°	30–35°	1.0	low (< 5% total)	d(GG), d(AG)	no
trans-DDP	9°	hinge joint	not recognized	low (< 5% total)	d(GNG) major	?
1,1/t,t	10°	—	0.3	high	d(GG)	yes
1,1/c,c	12°	—	0.15	high	not seen	yes

Notes: [a]Values for *cis*- and *trans*-DDP taken from References 52, 68, and 69. Values for dinuclear platinum complexes from Ref. 39.
[b]From Ref. 70.
[c]From Ref. 77.

cis-DDP is achievable but without the necessity of the *cis*-DDP structure. If unwinding is responsible for recognition of *cis*-DDP-damaged DNA,[70] then similar effects should be observed from the dinuclear complexes.

A principal feature of the *cis*-DDP intrastrand cross-link is a kink or bend of the double helix toward the major groove.[71] Recent studies have implicated this conformational change as being responsible for the recognition of *cis*-DDP-damaged DNA, but not of DNA damaged by *trans*-DDP or [PtCl(dien)]⁺, by the family of HMG proteins.[72,73] Damage recognition or structure-specific recognition proteins contain the HMG structural motif.[73,74] Attention has been drawn to similarities between the "natural" bending of DNA required for transcription and other cellular functions and the bending induced by DNA-damaging agents such as *cis*-DDP, which is almost certainly the recognition motif for the protein.[75]

Using a damaged DNA-affinity assay,[76] we explored the HMG recognition of DNA adducted by bis(Pt) complexes. HMG proteins recognize DNA damaged by both dinuclear compounds, but not as efficiently as it does *cis*-DDP-damaged DNA (Figure 8).[77] Quantitation of the binding of the dinuclear complexes to DNA showed that at equal r_b and equal concentrations of protein, recognition is 30% and 15% of that for *cis*-DDP for the 1,1/t,t and 1,1/c,c complexes, respectively. Further studies show that HMG recognition is also dependent on chain length within the 1,1/t,t series, where $n = 6 > 4 > 2$.[77] The fact that HMG proteins recognize dinuclear complexes with monofunctional coordina-

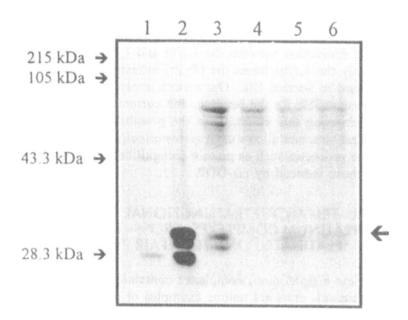

Figure 8. Binding of HMG proteins to DNA modified by *cis*-DDP and dinuclear platinum complexes. The damaged DNA affinity precipitation assay was performed with proteins extracted from CHO cell nuclei and DNA containing an equal number of drug adducts (r_b). The arrow on the right shows the position of the HMG proteins. Lane 1 shows molecular weight markers; lane 2, standard protein extract; lane 3, HMG protein binding to DNA damaged by *cis*-DDP; lane 4, protein binding to undamaged DNA; lane 5, HMG protein binding to DNA damaged by {[*trans*-PtCl(NH$_3$)$_2$]$_2$(H$_2$N(CH$_2$)$_4$NH$_2$}$^{2+}$ (1,1/t,t); and lane 6, HMG protein binding to DNA damaged by {[*cis*-PtCl(NH$_3$)$_2$]$_2$H$_2$N(CH$_2$)$_4$NH$_2$}$^{2+}$ (1,1/c,c). Selection of an arbitrary value of 100% for the binding to *cis*-DDP and quantitation of binding gave levels of 30% for 1,1/t,t and 15% for 1,1/c,c. The higher molecular weight proteins in lanes 3–6 originate from the milk solution used to block the nonspecific DNA binding sites. See Reference 78 for details.

tion spheres but *do not* recognize the analogous monomer [PtCl(dien)]Cl again confirms that the conformational changes induced by the dinuclear complexes are not simply the sum of two mononuclear or monofunctional lesions.[39]

The protein recognition thus implies that some bending similar to that induced by *cis*-DDP is also induced by the dinuclear compounds. Note that the two bis(platinum) isomers are not equally efficient with respect to HMG protein recognition. The critical distinction between the 1,1/t,t and 1,1/c,c complexes is that only the 1,1/t,t forms the (Pt,Pt) intrastrand cross-link, as explained in Section IIIC. Our results imply that this lesion is then responsible for the bending, and current studies are aimed at quantitating this effect. Thus the possibility arises that the dinuclear structure allows us to *systematically* modify important cellular processes such as protein recognition of damaged DNA from those induced by *cis*-DDP.

IV. TRI- AND TETRAFUNCTIONAL DINUCLEAR PLATINUM COMPLEXES; CROSS-LINKING OF PLATINATED DNA TO REPAIR PROTEINS

Dinuclear bis(platinum) complexes containing one or two *cis*-PtCl$_2$(amine)$_2$ units are unique examples of potentially trifunctional and tetrafunctional antitumor and DNA-binding agents (see Figure 2). In these cases, the second step of binding to DNA is important because a competition arises between formation of (Pt,Pt) interstrand cross-links or a *cis*-DDP-like intrastrand adduct by preferential reaction of one *cis*-Pt unit.[12,78] It is likely that *cis*-DDP-like binding will occur, and thus some conformational changes similar to the mononuclear complex may be expected.[16] Relevant to this point is the fact that the series [{*cis*-PtCl$_2$(NH$_3$)}$_2$H$_2$N(CH$_2$)$_n$NH$_2$] (2,2/c,c) does not induce the B–Z transition in poly(dG–dC)·poly(dG–dC). Nevertheless, the possibility of trifunctional and tetrafunctional binding allows the possibility of formation of novel DNA adducts.

The excision repair UvrABC complex recognizes bis(Pt)–DNA adducts formed by either tetrafunctional or bifunctional complexes.[14,16,22] The availability of multiple coordination sites (> 2) in the former suggested that covalent cross-linking of a DNA binding protein to platinated DNA could occur. This possibility was recently confirmed for both homodinuclear (Pt,Pt) and heterodinuclear (Ru,Pt) complexes using the radio-labeled 49-bp DNA fragment, described previously, combined with native and denaturing polyacrylamide gel electrophoresis. A radiolabeled bis(Pt)–DNA fragment was mixed with UvrA and the products of the reactions were heated in a buffer containing SDS (which denatures proteins and causes any noncovalently bound protein to dissociate) and were analyzed by electrophoresis on a denaturing polyacrylamide gel.[79] The radiolabeled DNA became irreversibly bound to the UvrA protein in a protein-concentration- dependent manner, as evidenced by the co-migration of the radiolabeled DNA fragment with the protein producing a radioactive band at the site of where UvrA (100 kdaltons) migrates.

The DNA lesion responsible for efficient protein–DNA cross-linking is most probably a DNA–DNA interstrand cross-link in which each metal atom is coordinated with one strand of the DNA helix (Figure 9). The cross-linking efficiency is significantly greater than for mononuclear complexes such as *cis-* and *trans*-DDP and suggests novel ways for design of specific DNA–

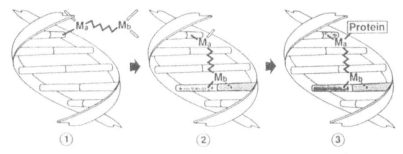

Figure 9. Scheme for formation of ternary Pt-DNA–protein cross-linking. Cross-linking of UvrAB proteins probably occurs through a pre-formed DNA–DNA interstrand cross-link (1 → 2 → 3). This mode of DNA–protein cross-linking is only available through a polyfunctional complex such as a dinuclear species. See Reference 14.

protein cross-linking agents. The formation of ternary DNA–protein adducts could have significance in developing "suicide" agents capable of inactivating repair proteins as well as in isolation of DNA-bound proteins from cells.

V. SUMMARY

Bifunctional binding of platinum complexes to DNA is considered essential for manifestation of antitumor activity. The studies on $[\{PtCl(NH_3)_2\}_2H_2N(CH_2)_4NH_2]^{2+}$ show that complexes structurally different to cis-DDP and acting on DNA in a manner distinct from the clinically used agent may display an altered spectrum of antitumor activity. Table 2 summarizes how bifunctional binding may be systematically modified by bis(platinum) complexes to give lesions that are structurally distinct from the cis-DDP–DNA adducts, thus affecting both conformational changes and sequence specificity.

Inhibition of both DNA synthesis and transcription has been implicated as the locus of cisplatin action.[28,80,81] The clinical efficacy of cis-DDP may be eventually due to differential repair of the lesions formed, a point that is also important in development of cis-DDP resistance. The location of adducts within specific genes is also expected to influence these factors to a greater extent than formation of adducts in nontranscribing regions of DNA.[46] Randomly platinated DNA is a substrate for the UVrABC repair system,[67,82] but individual adducts are not all excised with equal efficiency.[83] Emphasizing that the structurally different lesions of cis-DDP produce different cellular responses is the enhanced mutagenicity of the d(ApG) adduct in comparison to the d(GpG) adduct.[84,85] Likewise, the structural features of the bis(platinum)–DNA adducts will be expected to have further biological consequences such as inhibition of DNA synthesis or altered susceptibility to DNA repair. The altered profile of Pt–DNA binding described here confirms that useful biological activity is not restricted to direct analogues of the cis-DDP structure. The potential for systematic variation of DNA-binding properties within the dinuclear structure affords the promise for rational drug design and more specific inhibition of critical cellular and biochemical targets.

ACKNOWLEDGMENTS

I thank my many co-workers for the results presented. This work is supported by The American Cancer Society (Grant DHP-2E) and Boehringer Mannheim Italia.

REFERENCES

1. *Cancer Chemotherapy, Principles and Practice;* Chabner, B. A., Collins, J. M., Eds.; Lippincott: Philadelphia, 1990, pp. 465–490.
2. Rosenberg, B. In *Nucleic Acid–Metal Ion Interactions,* Vol. 1; Spiro, T. G., Ed.; Wiley: New York, 1980, 1–29.
3. Sherman, S. E.; Lippard, S. *J. Chem. Rev.* **1987**, *87,* 1153–1181.
4. Reedijk, J.; Fichtinger-Schepman, A. M. J.; van Oosterom, A. T.; van de Putte, P. *Structure and Bonding* **1987**, *67,* 53–89.
5. Bohr, V. A. *Carcinogenesis* **1991**, *12,* 1983–1992.
6. Chu, G.; Chang, E. *Science* **1988**, *242,* 564–567.
7. Christian, M. C. *Seminars in Oncology* **1992**, *19,* 720–733.
8. Eisenhauer, E.; Swerton, K.; Sturgeon, J.; Fine, S.; O'Reilly, S.; Canetta, R. In *Carboplatin: Current Perspectives and Future Directions;* Bunn, P., Canetta, R., Ozols, R., Rozencweig, M., Eds.; W. B. Saunders: Philadelphia, pp. 133–140.
9. Hoeschele, J. D.; Kraker, A. J.; Qu, Y.; Van Houten, B.; Farrell, N. In *Molecular Basis of Specificity in Nucleic Acid–Drug Interactions;* Pullman, B., Jortner, J., Eds.; Kluwer Academic Press: Dordrecht, 1990, 301–321.
10. Farrell, N.; Qu, Y.; Hacker, M. P. *J. Med. Chem.* **1990**, *33,* 2179–2184.
11. Farrell, N. *Cancer Investigation* **1993**, *11,* 578–579.
12. Farrell, N. *Comments in Inorganic Chemistry* **1995**, *16,* 373–389.
13. Qu, Y.; Farrell, N. *Inorg. Chem.* **1995**, *16,* 373–384.
14. Van Houten, B.; Illenye, S.; Qu, Y.; Farrell, N. *Biochemistry* **1993**, *32,* 11794–11801.
15. Manzotti, C.; Pezzoni, G.; Giuliani, F.; Valsecchi, M.; Farrell, N.; Tognella, S. *Proc. Am. Assoc. Cancer Res.* **1994**, *35,* 2628.
16. Farrell, N.; Qu, Y.; Feng, L.; Van Houten, B. *Biochemistry* **1990**, *29,* 9522–9531.
17. Farrell, N. P.; de Almeida, S. G.; Skov, K. A. *J. Am. Chem. Soc.* **1988**, *110,* 5018–5019.
18. Wu, P. K.; Qu, Y.; Van Houten, B.; Farrell, N. *J. Inorg. Biochem.* **1994**, *54,* 207–220.
19. Chu, G. Y. H.; Mansy, S.; Duncan, R. F.; Tobias, R. S. *J. Am. Chem. Soc.* **1978**, *100,* 593–606.
20. Zou, Y.; Van Houten, B.; Farrell, N. *Biochemistry* **1993**, *32,* 9632–9638.

21. Perez, R. P.; Farrell, N.; Hamilton, T. C. *Proc. Am. Assoc. Cancer Res.* **1993**, *34*, 2395.
22. Roberts, J. D.; Van Houten, B.; Qu, Y.; Farrell, N. P. *Nucleic Acids Res.* **1989**, *17*, 9719–9733.
23. Gruff, E. S.; Orgel, L. E. *Nucleic Acids Res.* **1991**, *19*, 6849–6857.
24. Hartley, J. A.; Souhami, R. L.; Berardini, M. D. *J. Chromatography* **1993**, *618*, 277–288.
25. Weidner, M. F.; Millard, J. T.; Hopkins, P. B. *J. Am. Chem. Soc.* **1989**, *111*, 9270–9272.
26. Millard, J. T.; Weidner, M. F.; Kirchner, J. J.; Ribeiro, S.; Hopkins, P. B. *Nucleic Acids Res.* **1991**, *19*, 1885–1891.
27. Schwartz, A.; Sip, M.; Leng, M. *J. Am. Chem. Soc.* **1990**, *112*, 3673–3674.
28. Lemaire, M.-A.; Schwartz, A.; Rahmouni, A. R.; Leng, M. *Proc. Natl. Acad. Sci. U.S.A.* **1991**, *88*, 1982–1985.
29. Zou, Y.; Van Houten, B.; Farrell, N. *Biochemistry* **1994**, *33*, 5404–5410.
30. Royer-Pokora, B.; Gordon, L. K.; Haseltine, W. A. *Nucleic Acids Res.* **1981**, *9*, 4595–4609.
31. Fuchs, R. P. P.; Koffel-Schwartz, N.; Daune, M. P. In *Cellular Responses to DNA Damage;* Alan R. Liss: New York, 1983, pp. 547–557.
32. Malinge, J.-M.; Schwartz, A.; Leng, M. *Nucleic Acids Res.* **1987**, *15*, 1779–1797.
33. Hemminki, K.; Thilly, W. G. *Mutat. Res.* **1988**, *202*, 133–138.
34. Marrot, L.; Leng, M. *Biochemistry* **1989**, *28*, 1454–1461.
35. Gralla, J. D.; Sasse, D. S.; Poljak, L. G. *Cancer Res.* **1987**, *47*, 5092–5096.
36. Buchanan, R. L.; Gralla, J. D. *Biochemistry* **1990**, *29*, 3436–3442.
37. Lempers, E. L. M.; Bloemink, M. J.; Reedijk, J. *Inorg. Chem.* **1991**, *30*, 201–211.
38. van der Veer, J. L.; van der Marel, G. A.; van den Elst, H.; Reedijk, J. *Inorg. Chem.* **1987**, *26*, 2272–2279.
39. Farrell, N.; Appleton, T. G.; Qu, Y.; Roberts, J. D.; Soares Fontes, A. P.; Skov, K. A.; Wu, P. ; Zou, Y. *Biochemistry* **1995**, *34*, 15480–15486.
40. Bloemink, M. J.; Reedijk, J.; Farrell, N.; Qu, Y.; Stetsenko, A. I. *J. Chem. Soc., Chem. Commun.* **1992**, 1002–1003.
41. den Hartog, J. H. J.; Altona, C.; Chottard, J.-C. *Nucleic Acids Res.* **1982**, *10*, 4715–4730.
42. Sherman, S. E.; Gibson, D.; Wang, A. H. J.; Lippard, S. J. *Science* **1985**, *230*, 412–417.
43. Qu, Y.; Bloemink, M. J.; Reedijk, J.; Farrell, N. Manuscript in preparation.
44. Inagaki, K.; Kidani, Y. *Inorg. Chem.* **1986**, *25*, 1–9.
45. van Garderen, C. J.; van Houte, L. P. A.; van den Elst, H.; van Boom, J. H.; Reedijk, J. *J. Am. Chem. Soc.* **1989**, *111*, 4123–4125.
46. Jones, J. C.; Zhen, W.; Reed, E.; Parker, J. R.; Sancar, A.; Bohr, V. A. *J. Biol. Chem.* **1991**, *266*, 7101–7107.
47. Syzmkowski, D. E.; Yarema, K.; Essigmann, J. M.; Lippard, S. J.; Wood, R. D. *Proc. Natl. Acad. Sci. U.S.A.* **1992**, *89*, 10772–10776.
48. Richon, W. M.; Schulte, N. A.; Eastman, A. *Cancer Res.* **1987**, *47*, 2056–2061.
49. van Garderen, C. J.; Altona, C.; Reedijk, J. *Inorg. Chem.* **1990**, *29*, 1481–1487.

50. Brabec, V.; Reedijk, J.; Leng, M. *Biochemistry* **1992**, *31*, 12397–12402.
51. Johnson, A.; Illenye, S.; Farrell, N.; Van Houten, B. *Proc. AACR,* **1996**, *35*, 2637.
52. Keck, M. V.; Lippard, S. J. *J. Am. Chem. Soc.* **1992**, *114*, 3386–3390.
53. Johnson, A.; Qu, Y.; Van Houten, B.; Farrell, N. *Nucleic Acids Res.* **1992**, *20*, 1697–1703.
54. Malfoy, B.; Hartmann, B.; Leng, M. *Nucleic Acids Res.* **1981**, *9*, 5659–5669.
55. Ushay, M.; Santella, R. M.; Grunberger, D.; Lippard, S. J. *Nucleic Acids Res.* **1982**, *10*, 3573–3588.
56. Gessner, R.V.; Quigley, G. J.; Wang, A. H. J.; van der Marel, G. A.; van Boom, J. H.; Rich, A. *Biochemistry* **1985**, *24*, 237–240.
57. Ho, P. S.; Frederick, C. A.; Saal, D.; Wang, A. H. J.; Rich, A. *J. Biomolec. Struct. Dynam.* **1987**, *4*, 521–534.
58. Qu, Y.; Farrell, N. *Inorg. Chim. Acta* **1996**, *245*, 265–267.
59. Farrell, N.; Qu, Y. *Inorg. Chem.* **1989**, *28*, 3416–3420.
60. Behe, M.; Felsenfeld, G. *Proc. Natl. Acad. Sci. U.S.A.* **1981**, *78*, 1619–1623.
61. Rich, A.; Nordheim, A.; Wang, A. H. J. *Ann. Rev. Biochem.* **1984**, *53*, 791–846.
62. Farrell, N.; Wu, P. Unpublished results.
63. Feuerstein, B. G.; Williams, L. D.; Basu, H. S.; Marton, L. J. *J. Cell Biochem.* **1991**, *46*, 37–47.
64. Castleman, H.; Hanau, L. H.; Erlanger, B. F. *Nucleic Acids Res.* **1983**, *23*, 8421–8429.
65. Liu, L. F.; Wang, J. C. *Proc. Natl. Acad. Sci. U.S.A.* **1987**, *84*, 7024–7027.
66. Rich, A. In *Proceedings of The Robert A. Welch Foundation*, Vol. 37; Houston, 1993, pp. 13–34.
67. Sancar, A.; Sancar, G. B. *Ann. Rev. Biochem.* **1988**, *57*, 29–67.
68. Cohen, G. L.; Bauer, W. R.; Barton, J. K.; Lippard, S. J. *Science* **1979**, *203*, 1014–1016.
69. Scovell, W. M., Collart, F. *Nucleic Acids Res.* **1985**, *13*, 2881–2895.
70. Bellon, S. F.; Coleman, J. H.; Lippard, S. J. *Biochemistry* **1991**, *30*, 8026–8035.
71. Rice, J. A.; Crothers, D. M.; Pinto, A. L.; Lippard, S. J. *Proc. Natl. Acad. Sci. U.S.A.* **1988**, *85*, 4158–4161.
72. Billings, P. C.; Davis, R. J.; Engelsberg, B. N.; Skov, K. A.; Hughes, E. N. *Biochem. Biophys. Res. Commun.* **1992**, *188*, 1286–1294.
73. Pil, P. M.; Lippard, S. J. *Science* **1992**, *256*, 234–237.
74. Brown, S. J.; Kellett, P. J.; Lippard, S. J. *Science* **1993**, *261*, 603–605.
75. Lilley, D. M. J. *Nature* **1992**, *357*, 282–283.
76. Marples, B.; Adomat, H.; Billings, P. C.; Farrell, N. P.; Koch, C. J.; Skov, K. A. *Anti-Cancer Drug Design* **1994**, *9*, 389–399.
77. Skov, K. A.; Adomat, H.; Farrell, N.; Marples, B.; Matthews, J.; Walter, P.; Qu, Y.; Zhou, H. *Proc. Am. Assoc. Cancer Res.* **1993**, *34*, 2571.
78. Qu, Y.; Farrell, N. *J. Am. Chem. Soc.* **1991**, *113*, 4851–4857.
79. Laemmli, U. K. *Nature* **1970**, *227*, 680–685.
80. Roberts, J. J.; Pera, M. F. In *Molecular Aspects of Anti-Cancer Drug Action;* Neidle, S., Waring, M. J., Eds.; Macmillan: London, 1983, pp. 183–231.
81. Eastman, A. *Pharmacol. Ther.* **1987**, *34*, 155–166.

82. Van Houten, B. *Microbiological Review* **1990**, *54*, 18–51.
83. Page, J. D.; Husain, I.; Sancar, A.; Chaney, S. G. *Biochemistry* **1990**, *29*, 1016–1024.
84. Burnouf, D.; Daune, M.; Fuchs, R. P. P. *Proc. Natl. Acad. Sci. U.S.A.* **1987**, *84*, 3758–3762.
85. Naser, L. J.; Pinto, A. L.; Lippard, S. J.; Essigmann, J. M. *Biochemistry* **1988**, *27* 4357–4367.

DNA SEQUENCE SELECTIVITY OF THE PYRROLE-DERIVED, BIFUNCTIONAL ALKYLATING AGENTS

Paul B. Hopkins

Advances in DNA Sequence Specific Agents
Volume 2, pages 217–239.
Copyright © 1996 by JAI Press Inc.
All rights of reproduction in any form reserved.
ISBN: 1-55938-166-3

ABSTRACT

Several of the substances used in the treatment of cancer in humans are bifunctional alkylating agents. The biological target of these substances is believed to be DNA; interstrand or intrastrand DNA–DNA cross-linking is likely to be important in this regard. This chapter describes the DNA–DNA cross-linking reactions of a family of bifunctional electrophiles that show remarkable similarity with regard to the nucleotide sequence at which they form interstrand cross-links [5'-d(CG)] and the atomic sites that are linked (N-2 of deoxyguanosines on opposing strands). These substances include reductively activated mitomycin C, oxidatively activated pyrrolizidine alkaloids, 2,3- and 3,4-bis(acetoxymethyl)-1-methylpyrrole, a distamycin conjugate of 2,3-bis(hydroxymethyl)-1-methylpyrrole, FR900482, and FR66979. The feature that links these structurally diverse substances to one another is the presence of, or the ability to form, a pyrrole ring substituted at adjacent carbons with masked electrophilic centers. This geometry is an excellent match to the nucleophilic sites on N-2 of deoxyguanosines at the duplex sequence 5'-d(CG), possibly accounting for some or all of the atomic site and sequence specificities of these substances.

I. INTRODUCTION

The mutagenic, carcinogenic, and antitumor activities of DNA-alkylating agents have long been recognized, prompting scientists to devote considerable attention to unraveling the details of the alkylation reactions of DNA.[1,2] Issues of selectivity have from the outset figured prominently in this field; however, the *types* of selectivity amenable to study have evolved through the years. Early on, the focus of research into the chemistry of DNA alkylation was limited to defining the identity of the nucleotide and the site on that nucleotide targeted by an alkylation reaction. The development of DNA-sequencing and synthesis technologies now permits these same issues to be addressed in the context of surrounding nucleotide sequence. Definition of the selectivities expressed by the biologically active alkylating agents and elucidation of their mechanistic origins are contemporary activities.

One subset of the DNA alkylating agents is those that are bifunctional. This group includes several important antitumor substances, such as cisplatin, the nitrogen mustards, nitrosoureas, and mitomycin C. The idea that cross-linking of biopolymers,[3] and in particular nucleic acids,[4] might be important in explaining the cytological effects of the bifunctional alkylating agents is an old one. Many lines of evidence indicate that alkylation of DNA underlies these biological effects (see, for example, the case of nitrogen mustard).[5] Of particular interest is the observation that these substances can interact with DNA to form DNA–DNA interstrand cross-links,[6] because these linkages inhibit the strand separation of duplex DNA[7] expected to be critical for replication or transcription of DNA. Many studies have in fact positively correlated the cytotoxicity of bifunctional alkylating agents with their ability to form DNA–DNA interstrand cross-links,[8–10] although this has not always been so.[11,12]

The study of issues of selectivity in DNA interstrand cross-linking reactions of bifunctional alkylating agents has lagged behind the corresponding studies of monofunctional agents. The determination of the sites of covalent attachment between cross-linking agent and DNA has often proven challenging, because DNA–DNA cross-linking reactions are in many cases of low efficiency, producing a multicomponent mixture of structurally complex products. Techniques that have been employed to great advantage to reveal nucleotide sequence specificities of monoalkylating agents[13,14] have proven inapplicable to most interstrand cross-linked DNAs, because they do not undergo the strand separation and/or strand cleavage reactions, which are a critical step. The development of new techniques has in many cases been required to approach these problems.

The focus of this chapter is a family of DNA interstrand cross-linking agents, which on superficial inspection bear only limited structural resemblance to one another (Figure 1). These agents include mitomycin C (1), a clinically useful antitumor substance first isolated from *Streptomyces caespitosus;*[15,16] retrorsine (2), a member of the widely distributed pyrrolizidine alkaloid family, members of which are commonly hepatotoxins and carcinogens;[17] the synthetic distamycin–pyrrole conjugate 3, designed and synthesized to test some of the hypotheses developed

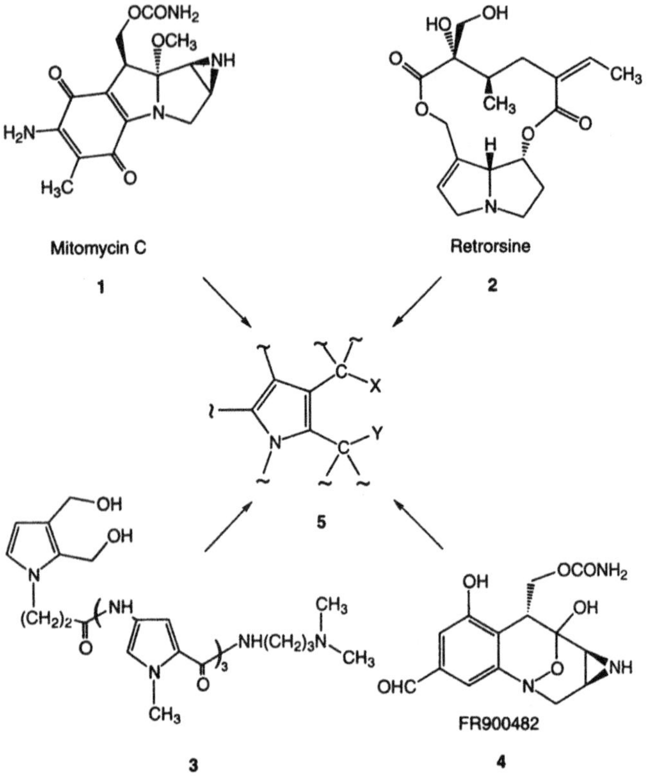

Figure 1. Structures of DNA interstrand cross-linking agents.

herein;[18] and FR900482 (**4**),[19] an antibiotic antitumor substance relatively recently isolated from *Streptomyces sandaensis*. The case will be made that what these substances and several others as well have in common is the presence of (as in **3**), or the ability to form (as in **1**, **2**, and **4**), a disubstituted pyrrole here represented as **5**. The replacement of groups X and Y of **5** (Figure 2) by nucleophilic groups in DNA is the basis of their bifunctional alkylation reactions. Remarkably, all of these substances preferentially interstrand cross-link the same dinucleotide sequence in duplex DNA, 5'-d(CG). Available evidence indicates that in all examples the site of alkylation is the same, N-2 of the deoxyguanosine residues on opposite strands. Although the

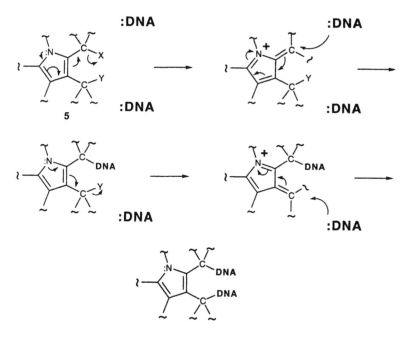

Figure 2. Possible mechanistic sequence for DNA–DNA cross-linking by a pyrrole-derived, bifunctional alkylating agent.

mechanistic details remain obscure, the pyrrole substructure **5** appears to be an excellent match to the structure of DNA at the cross-linked sequence.

The alkylation chemistry of a number of DNA–DNA inter-strand cross-linking reactions has been elucidated in recent years. The interested reader is referred elsewhere for descriptions of cross-linking reactions of azinomycin,[20] butadiene bis-epoxide,[21] cisplatin,[22–24] diaziridinyl quinones,[25] formaldehyde,[26] nitrogen mustard,[27,28] and nitrous acid.[29]

II. THE COMPLEX CASE OF MITOMYCIN C

Mitomycin C (**1**) has been widely used in the treatment of cancer. Because several excellent reviews have appeared,[15,16] only information relevant to the theme of pyrrole-derived cross-linking agents is reviewed here.

Soon after the discovery of the mitomycins, it was recognized
that DNA is their likely target. Mitomycin C itself and DNA
do not react with one another at an appreciable rate at pH 7.
However, in the presence of a reducing agent, DNA alkylation
reactions occur, DNA–DNA interstrand cross-linking being
among them. Iyer and Szybalski[30] proposed in 1964 the fundamental elements of the ensuing chemical reactions. The reduction of the quinone ring facilitates loss of the elements of
methanol, generating a pyrrole ring. Sequential departure of the
leaving groups at C-1 and C-10 is thus facilitated, because the
resulting electron-deficient centers are conjugated with the
newly formed pyrrole ring (Figure 3). Nucleophilic attack of

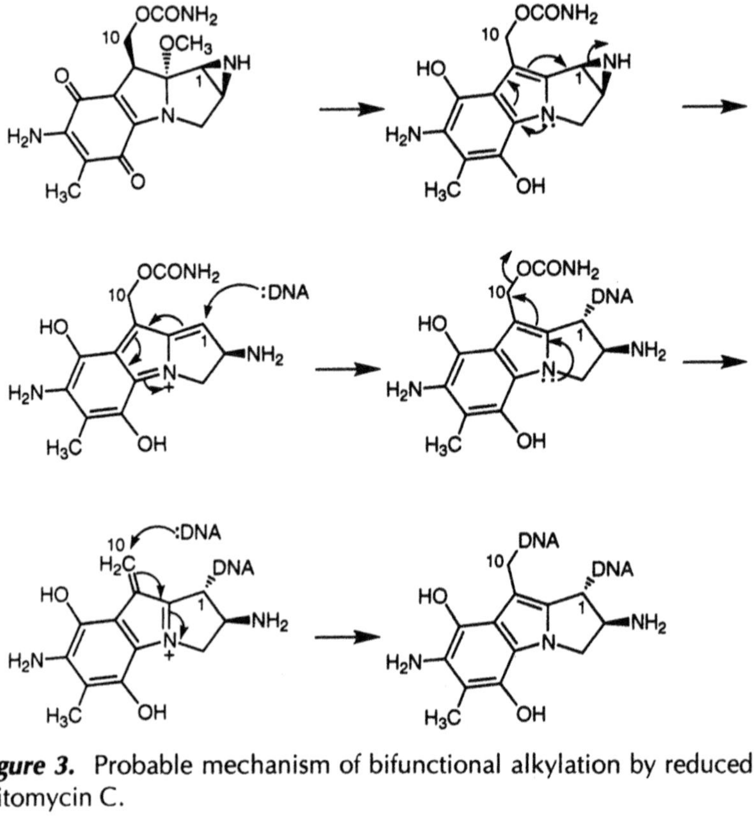

Figure 3. Probable mechanism of bifunctional alkylation by reduced
mitomycin C.

6

DNA at these centers results in the bifunctional alkylation of DNA. The question of whether some or all of these intermediates should be depicted as semiquinone radicals remains open. The first of the two sites on DNA alkylated by reductively activated mitomycin C was identified as N-2 of deoxyguanosine.[31]

The report that sparked our interest in this area was that of Tomasz and co-workers, in which the second site was conclusively identified as also being N-2 of deoxyguanosine.[32] Enzymatic hydrolysis of the phosphodiester backbone of DNA treated with mitomycin C (and with a reducing agent for in vitro experiments) afforded the conjugate **6** of two moles of deoxyguanosine and one of mitomycin. The structure of this substance, available at that time in minute quantities, was carefully characterized as its peracetylated derivative in an elegant series of spectroscopic measurements. If this substance was derived from the nucleus of the interstrand cross-link, it seemed probable that the lesion had been formed at either or both of the DNA sequences 5'-d(CG) and 5'-d(GC). These sequences would place the nucleophilic centers of the two deoxyguanosine residues with a spacing that approximately matches that of the electrophilic target sites in activated mitomycin C.

The sequence specificity of DNA–DNA interstrand cross-linking by reductively activated mitomycin C has been tested in three different experiments, all of which support the same conclusion. In one experiment, short synthetic DNA duplexes containing either a single 5'-d(CG) site, a single 5'-d(GC) site, or a single 5'-d(GG) site were incubated with reductively activated mitomycin C. Digestion of the phosphodiester backbone afforded **6** only for the 5'-d(CG)- and 5'-d(GG)-containing DNAs, indicating that interstrand and intrastrand lesions could be formed at these two sequences, respectively.[33,34] In a separate experiment, synthetic DNA duplexes containing various sequences were incubated with reductively activated mitomycin C. Denaturing polyacrylamide gel electrophoresis (DPAGE), which easily distinguishes single-strand from double-strand (cross-linked) DNA, then revealed that only those DNAs containing the sequence 5'-d(CG) formed interstrand cross-links.[35] Finally, the cross-linked nucleotides in mitomycin-treated DNA have been determined directly by a strand cleavage experiment. In this experiment, the cross-linked DNA, with one of the two strands radiolabeled at a defined terminus, is fragmented sequence-randomly and under single-hit conditions using a chemical nuclease. DPAGE analysis of the resulting fragment mixture then reveals the site of cross-linking on the radiolabeled strand, because radiolabeled fragments shorter than full-length single strand are formed only by cleavage reactions between the radiolabel and the interstrand cross-link.[36] By this method, deoxyguanosine at the 5'-d(CG) sequence was found to be linked to the opposite strand.[36,37] Taken together, these experiments conclusively prove that lesion 6 is formed at the sequence 5'-d(CG) and not at 5'-d(GC). Recently, a closely related lesion has been identified as the nucleus of the intrastrand cross-link.[38]

What is the mechanistic origin of this sequence specificity? Two observations are important in this regard. It has been shown that monoadducts of mitomycin C are formed approximately 10-fold more efficiently at 5'-d(CG) sites than at 5'-d(GC) sites.[39,40] Furthermore, monoadducts that are formed at 5'-d(GC) sites have been shown not to progress to cross-links, whereas those at 5'-d(CG) sites do so essentially quantitatively.[34] With regard to the selective monoadduct formation, experiments employing deoxyinosine provide strong evidence that monoalkylation is facilitated by the presence of an amino group to the 5'-side of the deoxyguanosine being alkylated.[39] At 5'-d(CG) sites, this group is provided by the deoxyguanosine on the opposite strand. This amino group has been proposed to hydrogen bond to the carbamoyl group, facilitating the first alkylation.[39]

The progression of monoadducts at 5'-d(CG) but not 5'-d(GC) to interstrand cross-links may have its origin in at least two factors. The first of these is what appears on the basis of model building,[31] computation,[37] and most recently solution NMR measurements[41] to be a strong tendency for the mitomycin nucleus in the monoadduct to align its long axis with the minor groove of DNA, and with an orientational preference for the carbamoyl function to point toward the 5'-end of the alkylated strand. An argument has been made that the chirality of the mitomycin nucleus at C-1 is important in this regard, having analogy in the orientation preference of the pyrrolobenzodiazepine (anthramycin) family.[37] This alone cannot be responsible, because, as shown in Section IV, related *achiral* agents retain a preference for interstrand cross-linking at 5'-d(CG).

It has been argued that structural differences in the relative orientation of the N-2 amino groups of deoxyguanosine at the sequences 5'-d(CG) and 5'-d(GC) may be important in this regard (Figure 4).[24,37,42] Molecular models of the interstrand crosslink with a covalent structure as in **6** at these two sequences clearly reveal that the 5'-d(CG) sequence can accommodate the linkage with minimal reorganization of the relative orientation of the two guanyl moieties, as is seen in the solution structure of a synthetic duplex interstrand cross-linked at this sequence

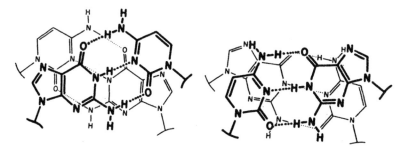

Figure 4. Geometry of 5'-d(CG) (left) and 5'-d(GC) (right) in Arnott B-DNA. The N-2 exocyclic amino groups of the two deoxyguanosines reside centrally in the minor groove, which appears at the bottom in these views down the helix axis.

with mitomycin C.[43] In contrast, the 5'-d(GC) sequence would be expected to require structural reorganization, as a consequence of the constraint imposed by the geometry of the mitosene skeleton on the orientation of the potentially nucleophilic centers in DNA. At 5'-d(CG), these sites are aligned, a good match to the mitosene skeleton, whereas at 5'-d(GC), they diverge, a geometry not allowed by the mitosene skeleton. The argument, then, is that the transition state for monoadduct progression to cross-link that is closest to the ground-state geometry of duplex DNA will be energetically favored, and thus that this reaction at 5'-d(CG) will be preferred.[24]

Inspection of molecular models suggests that, if the preceding hypothesis of minimal reorganization is valid, molecules simpler than mitomycin will preferentially cross-link 5'-d(CG). The quinone ring, for example, does not influence the relative geometry of the electrophilic centers. By chance, a critical test of this notion was conveniently available, as discussed in the following section.

III. THE PYRROLIZIDINE ALKALOID CONNECTION

Pyrrolizidine alkaloids are present in a wide variety of plant species around the world.[17] Their toxicity is believed to account for numerous livestock poisoning deaths. The contamination of various foodstuffs, such as honey and milk, by pyrrolizidine alkaloids is of concern, given their toxic, mutagenic, and carcinogenic activities. A partial structural overlap of the pyrrolizidine alkaloids, such as retrorsine (2), retronecine (7), and monocrotaline (8), and the mitomycins was noted many years ago.[30,44,45] This connection has grown closer as the biotransformations of all of these substances have become better understood.

Pyrrolizidine alkaloids undergo hepatic oxidation to highly reactive dehydro derivatives [e.g., dehydroretrorsine (9), dehydroretronecine (10), and dehydromonocrotaline (11)], which are most likely responsible for their toxic effects.[17] Pyrrolizidine alkaloids are DNA interstrand cross-linking agents in vivo;[46] the oxidatively activated pyrrolizidines, the dihydropyrrolizines,

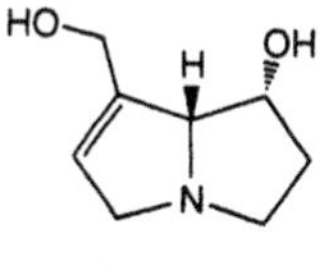

Retronecine

7

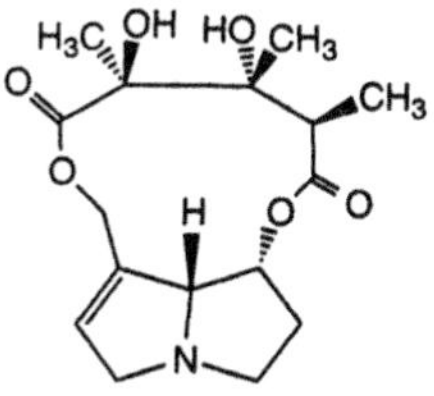

Monocrotaline

8

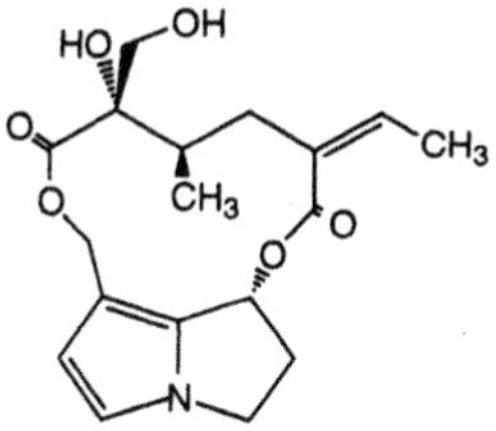

Dehydroretrorsine

9

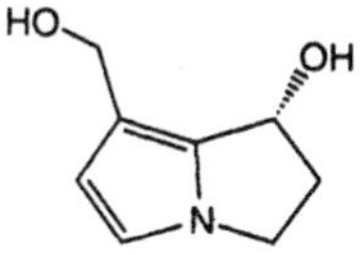

Dehydroretronecine

10

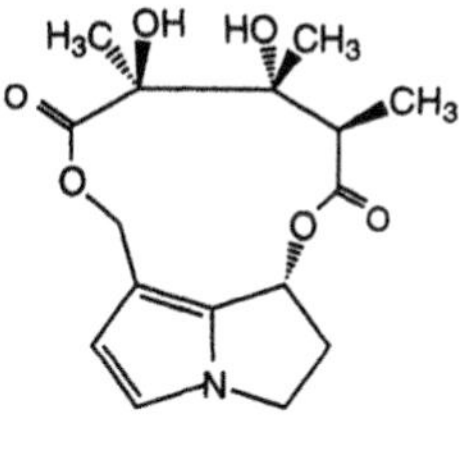

Dehydromonocrotaline

11

mimic this action in vitro.[47] The structural similarity of reductively activated mitomycins and oxidatively activated pyrrolizidines is striking (Figure 5). These agents possess the structural requirements for the reactions shown in Figure 2, suggesting that they may share common features in their DNA cross-linking reactions.

The DNA interstrand cross-linking reactions of dehydroretrorsine (**9**), dehydromonocrotaline (**11**), and the diacetate of dehydroretronecine (**10**) have been studied in synthetic oligonucleotide duplexes.[42] These agents were found to be very inefficient interstrand cross-linking agents, in comparison to reductively activated mitomycin C,[48] affording yields of under 1% in the DNAs studied. Clear evidence in favor of preferential

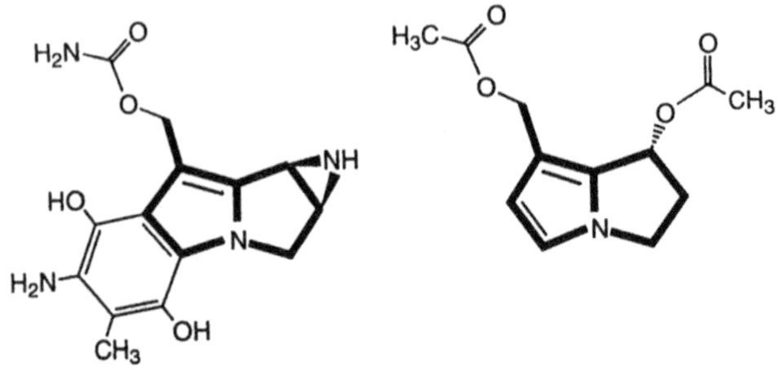

Figure 5. Structural overlap of reductively activated mitomycin C (a mitosene) with an oxidately activated pyrrolizidine alkaloid (a dehydropyrrolizine).

cross-linking at 5'-d(CG) rather than 5'-d(GC) was obtained when these sites were incorporated into distinct oligonucleotides and the products assayed by DPAGE. For the diacetate of dehydroretronecine, the random fragmentation method demonstrated that these cross-links bridge the deoxyguanosine residues at this sequence. Unlike the case of mitomycin C, it was possible to isolate a sufficient interstrand cross-link at the sequence 5'-d(GC) to analyze its connectivity by this same method, and it, too, was linked though the opposing deoxyguanosines. Obviously, the preference for 5'-d(CG) is not in this case absolute. For none of these compounds was the covalent connectivity of the interstrand cross-linked products determined.

The relationship of the reactions of mitomycin C and the pyrrolizidines was thus brought one step closer, as both substances were shown to cross-link the dinucleotide sequence 5'-d(CG) by bridging opposing deoxyguanosine residues. The desire to place this relationship on an even more intimate level prompted the studies described in the next section.

IV. THE ROLE OF THE PYRROLE

The substituted pyrroles **12** and **13** [2,3- and 3,4-bis(acetoxymethyl)-1-methylpyrrole (2,3- and 3,4-BAMP), respectively], particularly the former, can be considered to be very highly simplified structural analogues of activated mitomycins or ac-

2,3-BAMP   3,4-BAMP   IPP

12   **13**   **14**

tivated pyrrolizidines. They retain the critical substructure proposed in Figure 2 to be required for cross-linking reactions. All
other extraneous rings have been deleted, and the substances
are both achiral.

2,3-BAMP and 3,4-BAMP are both DNA interstrand cross-
linking agents in synthetic oligonucleotide duplexes.[42,49] Their
efficiencies are comparable to those observed with dihydropyrrolizines, or about 1% yield of interstrand cross-link.[42] In common with all of the other substances discussed thus far, they
cross-link the sequence 5'-d(CG) in preference to 5'-d(GC). Substitution of deoxyinosine for one or both deoxyguanosine residues at the duplex sequence 5'-d(CG) abolished interstrand
cross-linking, indicating that N-2 of deoxyguanosine is the
likely site of alkylation on both strands. The sequence-random
fragmentation protocol again showed that it was the deoxyguanosine residues at this sequence that were cross-linked.

15

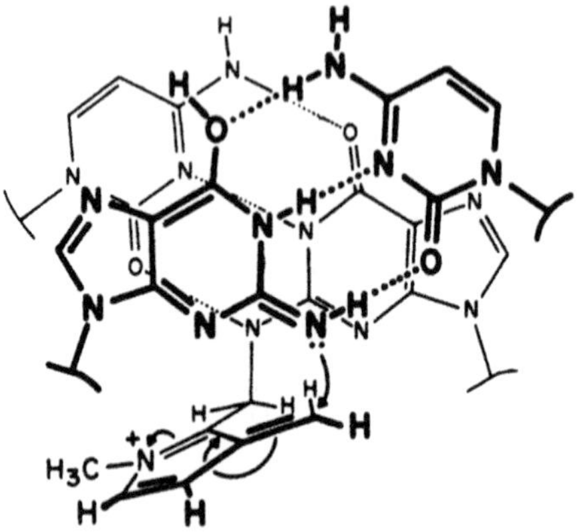

Figure 6. A possible geometry for the conversion of a monoadduct of 2,3-BAMP to an interstrand cross-link.

Attempts to determine the covalent structure of the interstrand cross-links derived from 2,3- and 3,4-BAMP were unsuccessful. Instead, a somewhat more complex, but still achiral substance, IPP (**14**), was employed.[49] This substance had many features in common with 2,3- and 3,4-BAMP, including being a DNA interstrand cross-linking agent that bridges deoxyguanosine residues preferentially at 5′-d(CG) and requires the N-2 amino group, but was distinct in one important respect: the yield of interstrand cross-link could be elevated to approximately 4%, thus allowing for more substantial structural characterization. Enzymatic hydrolysis of the phosphodiester backbone of a DNA duplex containing the sequence 5′-d(CG) interstrand cross-linked with IPP (**14**) afforded substance **15**, identified by comparison to an authentic sample prepared by rational chemical synthesis. This result established conclusively that the structural analogy of reductively activated mitomycin C (**1**) and IPP (**14**) extends to the covalent structures of their DNA interstrand cross-links.

Despite a reduction in absolute yield (approximately 50% for mitomycin C[48] vs. approximately 1% for **12** and **13**) and a reduction in absolute specificity for 5′-d(CG) over 5′-d(GC) (probably in excess of 100:1 for mitomycin C vs. approximately.

10:1 for **12** and **13**), the highly simplified analogues retain a preference for cross-linking the nucleotide sequence 5'-d(CG). As illustrated in Figure 6, it has been speculated that this may reflect the need for little reorganization of the geometry of DNA at this sequence in achieving the transition state for cross-linking.[42] This suggestion must be considered speculative, however, as it remains unknown whether or not these reactions are under kinetic control. Furthermore, it is unknown just how N-2 of deoxyguanosine serves as a nucleophile at all, given that the lone pair on this atom is delocalized in the π system of the attached purine and would be expected to be relatively inaccessible. Whether electrophiles attack this π-orbital directly, a rotated form of the amino group, or a tautomer (as shown in Figure 6) that exposes a lone pair in the minor groove, is at present unknown.

V. SEQUENCE-TARGETED PYRROLES

It is natural to ask whether or not the sequence specificities described in this and the companion contributions are biologically significant. It has been argued that given the elaborate design by which mitomycin ensures such specificity, there should be a significance to the observed sequence specificity.[50] Although this may be true for the antibiotic activity of mitomycin, where the low frequency of 5'-d(CG) sites in the mammalian relative to the bacterial genome is used to advantage, it is less obvious how this selectivity could account for antitumor activity. It is clear that the availability of agents that targeted even less common sites, or better yet, that could be designed to target specific sites such as those of an oncogene, would have consequences comparable to those promised by the antisense strategy.

2,3-BAMP (**12**) has been used as the "warhead" in a targeted interstrand cross-linking approach. Following the lead of Baker and Dervan, who have shown that conjugation of DNA-cleaving or alkylating agents to the minor-groove-binding agent distamycin affords so-called affinity cleavage reagents,[51] the substance **3** was designed and synthesized.[18] This substance was a remarkably efficient interstrand cross-linking agent, cross-linking a linearized plasmid at 10 nM. Comparable interstrand cross-linking with the analogue that lacked the distamycin function, 2,3-bis(hydroxymethyl)-1-methyl pyrrole, required a 1000-fold higher concentration of cross-linking agent. From use of synthetic oligonucleotide duplexes, it was shown that cross-linking occurs with a strong preference for sequences at which a distamycin-binding site, such as 5'-d(AAAA) or 5'-d(AATT), is joined directly to the preferred interstrand cross-linking site 5'-d(CG). Deoxyinosine substitution confirmed the requirement for N-2 of deoxyguanosine at both sites, and isolation of the conjugate afforded a substance with mass spectroscopic properties consistent with the expected structure. A duplex DNA containing the sequence 5'-d(GGAATT) was found to form an intrastrand, deoxyguanosine-to-deoxyguanosine cross-link in 33% yield. An intrastrand cross-link formed by reductively activated mitomycin C at the sequence 5'-d(GG) has recently been reported.[38] The toxicity of a homologue of **3** with one more methylene group in the tether linking distamycin to the pyrrole was evaluated in an L1210 cell line and showed an IC_{50} of 350 nM, a value comparable to that observed for the important antitumor agent cisplatin (Note 1).

VI.   FR66979 AND FR900482: MORE MASKED PYRROLES

A new group of antibiotic antitumor substances related in both structure and biological activity to the mitomycins has recently been reported. The naturally occurring FR900482 (**4**)[19] and FR66979 (**16**)[52] and a synthetic triacetate FK973 (**17**)[53] all possess antimicrobial and antineoplastic activity at approximately micromolar concentrations. FR900482 and FK973, like mitomycin C, produce DNA interstrand cross-links in vivo.[54] A time lag in the appearance of cross-links has been offered as evidence that these substances must undergo some form of activation in vivo.

The structural and biological similarity of the mitomycins to this new group of substances suggested that their mechanisms

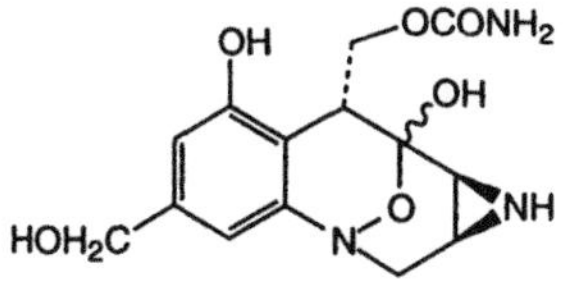

FR66979

16

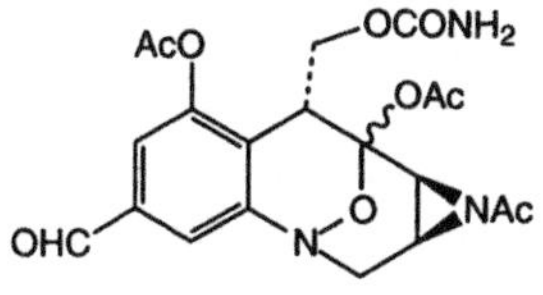

FK973

17

of action would share common features. However, the pyrrole function critical in the action of all of the substances discussed thus far is notably absent. Two solutions to this puzzle were proposed (Figure 7). In one, it was suggested that the N–O bond might be cleaved in vivo by reduction, exposing an amino ketone, which in turn might cyclize to the critical pyrrole **18**.[55] Alternatively, it was proposed that an S_N2'-displacement reaction followed by tautomerization might ultimately produce the closely related pyrrole **19**.[56] This amounts to an internal oxidation-reduction reaction. The suggestion that mitosene analogues might not be involved has been made.[57,58]

Studies of the reactions of FR900482 and FR66979 in synthetic oligonucleotide duplexes have shown their remarkable

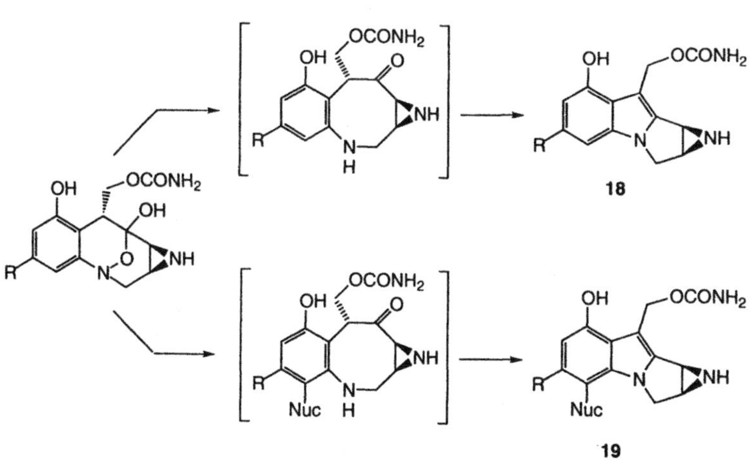

Figure 7. Proposed mechanisms for reductive (upper) and nucleophilic (lower) activation of FR900482 and FR66979.

similarity to mitomycin C.[59] No appreciable cross-linking is observed in the absence of a reducing agent, in this case sodium dithionite (Note 2). In the presence of dithionite, interstrand cross-links are formed efficiently only when the sequence 5'-d(CG) is present. The yield of interstrand cross-link is a function of flanking sequence, and for the three sequences studied the effect qualitatively parallels that seen with reductively activated mitomycin C. Deoxyinosine cannot substitute for deoxyguanosine, implicating N-2 of deoxyguanosine as the site of alkylation.

The lesion responsible for the interstrand cross-link was isolated from a synthetic DNA duplex containing a single 5'-d(CG) site interstrand cross-linked by dithionite-activated FR66979.[60] Peracetylation of the lesion was required to avoid an uncharacterized decomposition. The peracetylated derivative was unequivocally identified as 20 by a combination of UV, MS, and one- and two-dimensional ^{1}H-NMR measurements. This substance is the analogue of 6, the corresponding lesion from mitomycin C-treated DNA, and is most consistent with the intermediacy of the mitosene-like intermediate 18. Borohydride reduction of the corresponding lesion (without acetylation) from FR900482 yields a substance tentatively identified as the lesion 20 without its acetyl groups, suggesting that FR900482 undergoes an analogous activation and cross-linking reaction.[59]

In vitro experiments thus show the viability of a reductive pathway for activation of FR900482 and FR66979 to form DNA interstrand cross-linking agents similar to reductively activated mitomycin C. The possibility of a nucleophilic pathway remains a matter of speculation. The possibility of other metabolic trans-

20

formations should not at this stage be ruled out. For example, in principle the latent pyrrole ring could be exposed by *oxidation* of the phenol ring to a hydroquinone, which could in turn provide the reducing equivalents necessary for scission of the N–O bond of these substances leading to a pyrroloquinone. Only in vivo studies will reveal what is ultimately the biologically relevant pathway.

VII. CONCLUSIONS

Substances as structurally diverse as those in Figure 1 have thus been shown to have in common the remarkable property of identifying the same core sequence, 5'-d(CG), for interstrand cross-linking of duplex DNA. The presence of, or the ability to form, a pyrrole ring appears to be the feature that unites these disparate substances. This ring plays the dual role of being electron-rich, thus accelerating displacement reactions at the proximal carbons, and of orienting these electrophiles both in terms of interatomic spacing and atomic geometry in a way that is compatible with reaction with duplex DNA.

That molecules as small as some of those described herein are capable of displaying appreciable sequence discrimination in their cross-linking reactions may at first seem remarkable, given their modest size. Even *smaller* molecules display similar interstrand cross-linking sequence selectivities, such as cisplatin, which selects 5'-d(GC);[22–24] formaldehyde, which selects 5'-d(AT);[26] and nitrous acid, which selects 5'-d(CG), for interstrand cross-linking.[29] These reactions differ from those of the more widely studied monoalkylating agents in that the cross-linking agents must meet the rigid constraints of covalent bond length and angle at *two* sites on DNA rather than one. Because DNA displays in its ground-state structure pairs of nucleophilic sites in only a limited number of geometric relationships, and there is an energetic cost of reorganizing DNA, it is not surprising in hindsight that bifunctional alkylating agents of modest size might match the display of nucleophilic groups at only one or a small number of DNA sequences.

The sequence-discriminating abilities of the agents described herein are remarkable from the perspective of chemistry, but are woefully inadequate in the context of identifying one or a small number of sites in a genome. The expanded sequence preferences observed for the distamycin–pyrrole conjugate **3** suggest some future for the targeting of alkylating or cross-linking agents to a limited number of genomic sites. This approach will depend heavily on the outcome of efforts in other laboratories to develop agents that recognize by noncovalent means sequences of DNA 10 to 20 residues in length.

ACKNOWLEDGMENTS

It is a pleasure to acknowledge the efforts of the co-workers whose names are cited in the references and partial financial support from the National Institutes of Health (AG45804) and the State of Washington.

NOTES

1. We thank Dr. Paul Aristoff of The Upjohn Company for arranging the toxicity testing and for its interpretation.

2. In contrast to the results emphasized herein,[59,60] another group reported that FR66979 (**16**) produces DNA–DNA interstrand cross-links at 5'-d(CG)[58] in the *absence* of exogenous reducing agents.[57,58] The origin of this discrepancy may be the distinct sources of FR66979 employed by the two groups. In both cases, FR66979 was prepared by reduction of FR900482, but with different reducing agents. "Inactive" FR66979 was prepared by sodium borohydride reduction, whereas "active" (in the absence of exogenous reducing agent) FR66979 had its origin in hydrogenation on palladium. We tentatively propose that the latter preparation affords FR66979 but that it is contaminated by a further reduced substance of unspecified structure that is the active interstrand cross-linking agent. Because these studies employ FR66979 in large excess (typically a 200:1 ratio of agent to DNA),[58] this impurity at the approximate part per hundred level could account for *all* cross-linking. Presumably this same substance, or one closely related, is generated by the FR66979/sodium dithionite combination, thus accounting for the similar sequence specificity reported by the two groups.[58,59] The matter has recently been resolved in favor of this argument.[61] We thank Professor R. M. Williams for discussions of this matter.

REFERENCES

1. Singer, B. *Progress in Nucleic Acids Res. and Mol. Biol.* **1974**, *15*, 219–284.

2. Kohn, K. W. *Methods in Cancer Res.* **1979**, *16*, 291–345.
3. Goldacre, R. J.; Loveless, A.; Ross, W. C. J. *Nature* **1949**, *163*, 667–669.
4. Elmore, D. T.; Gulland, J. M.; Jordan, D. O.; Taylor, H. F. W. *Biochem. J.* **1948**, *42*, 308–316.
5. Gray, P.; Phillips, D. R. *Biochemistry* **1993**, *32*, 12471–12477.
6. Geiduschek, E. P. *Proc. Natl. Acad. Sci. U.S.A.* **1961**, *47*, 950–955.
7. Kohn, K. W.; Spears, C. L.; Doty, P. *J. Mol. Biol.* **1966**, *19*, 266–288.
8. O'Connor, P. M.; Kohn, K. W. *Cancer Commun.* **1990**, *2*, 387–394.
9. Zwelling, I. A.; Michaels, S.; Schwartz. H.; Dobson, P. P.; Kohn, K. W. *Cancer Res.* **1981**, *41*, 640–649.
10. Hansson, J.; Lewensohn, R.; Ringborg, U.; Nilsson, B. *Cancer Res.* **1987**, *47*, 2631–2637.
11. Kohn, K. W. In *Development of Target-Oriented Anticancer Drugs;* Cheng, Y.-C.; Goz, B.; Minkoff, M., Eds.; Raven Press: New York, 1983, pp. 181–187.
12. O'Conner, P. M.; Wasserman, K.; Sarang, M.; Magrath, I.; Bohr, W. A.; Kohn, K. W. *Cancer Res.* **1991**, *51*, 6550–6557.
13. Warpehoski, M. In *Advances in DNA Sequence Specific Agents.*, Vol. 1; Hurley, L. H., Ed.; JAI Press: Greenwich, 1992, pp. 217–245.
14. Kohn, K. W.; Hartley, J. A.; Mattes, W. B. *Nucleic Acids Res.* **1987**, *15*, 10531–10549.
15. Carter, S. K. In *Mitomycin C: Current Status and New Developments;* Carter, S. K., Crooke. S. T., Eds.; Academic Press: New York, 1958, pp. 251–254.
16. Tomasz, M. In *Topics in Molecular and Structural Biology: Molecular Aspects of Anticancer Drug–DNA Interactions,* Vol. 2; Neidle, S., Waring, M., Eds.; Macmillian: Basingstoke, U.K., 1994, pp. 312–349.
17. Mattocks, A. R. *Chemistry and Toxicology of Pyrrolizidine Alkaloids;* London: Academic Press, 1986.
18. Sigurdsson, S. Th.; Rink, S. M.; Hopkins, P. B. *J. Am. Chem. Soc.* **1993**, *115*, 12633–12634.
19. Uchida, S.; Takase, S.; Kayakiri, H.; Kiyoto, S.; Hashimoto, M. *J. Am. Chem. Soc.* **1987**, *109*, 4108–4109.
20. Armstrong, R. W.; Salveti, M. E.; Nguyen, M. *J. Am. Chem. Soc.* **1992**, *114*, 3144–3145.
21. Millard, J. T.; White, M. M. *Biochemistry* **1993**, *32*, 2120–2124.
22. Lemaire, M. S.; Schwartz, A.; Rahmouni, A. R.; Leng, M. *Proc. Natl. Acad. Sci. U.S.A.* **1991**, *88*, 1982–1985.
23. Zou, Y.; Van Houten, B.; Farrell, N. *Biochemistry* **1994**, *33*, 5404–5410.
24. Hopkins, P. B.; Millard, J. T.; Woo, J.; Weidner, M. F.; Kirchner, J. J.; Sigurdsson, S. Th.; Raucher, S. *Tetrahedron* **1991**, *47*, 2475–2489.
25. Alley, S. C.; Brameld, K. A.; Hopkins, P. B. *J. Am. Chem. Soc.* **1994**, *116*, 2734–2741.
26. Huang, H.; Hopkins, P. B. *J. Am. Chem. Soc.* **1993**, *115*, 9402–9408.
27. Rink, S. R.; Solomon, M. S.; Taylor, M. J.; Rajur, S. B.; McLaughlin, L. W.; Hopkins, P. B. *J. Am. Chem. Soc.* **1993**, *115*, 2551–2557.

28. Ojwang, J. O.; Grueneberg, D. A.; Loechler, E. L. *Cancer Res.* **1989**, *49*, 6529–6537.

29. Kirchner, J. J.; Sigurdsson, S. Th.; Hopkins, P. B. *J. Am. Chem. Soc.* **1992**, *114*, 4021–4027.

30. Iyer, V. N.; Szybalski, W. *Science* **1964**, *145*, 55–58.

31. Tomasz, M.; Chowdary, D.; Lipman, R.; Shimotakahara, S.; Veiro, D.; Walker, V.; Verdine, G. L. *Proc. Natl. Acad. Sci. U.S.A.* **1986**, *83*, 6702–6706.

32. Tomasz, M.; Lipman, R.; Chowdary, D.; Pawlak, J.; Verdine, G. L.; Nakanishi, K. *Science* **1987**, *235*, 1204–1208.

33. Chawla, A. K.; Lipman, R.; Tomasz, M. In *Structure and Expression, Volume 2: DNA and Its Drug Complexes;* Sarma, R. H., Sarma, M. D., Eds.; Academic Press: Albany, 1987, pp. 305–316.

34. Borowy-Borowski, H.; Lipman, R.; Tomasz, M. *Biochemistry* **1990**, *29*, 2999–3006.

35. Teng, S. P.; Woodson, S. A.; Crothers, D. M. *Biochemistry* **1989**, *28*, 3901–3907.

36. Weidner, M. F.; Millard, J. T.; Hopkins, P. B. *J. Am. Chem. Soc.* **1989**, *111*, 9270–9272.

37. Millard, J. T.; Weidner, M. F.; Raucher, S.; Hopkins, P. B. *J. Am. Chem. Soc.* **1990**, *112*, 3637–3641.

38. Bizanek, R.; McGuiness, B. F.; Nakanishi, K.; Tomasz, M. *Biochemistry* **1992**, *31*, 3084–3091.

39. Li, V.; Kohn, H. *J. Am. Chem. Soc.* **1991**, *113*, 275–283.

40. Kumar, S.; Lipman, R.; Tomasz, M. *Biochemistry* **1992**, *31*, 1399–1407.

41. Private communication, Prof. M. Tomasz.

42. Weidner, M. F.; Sigurdsson, S. Th.; Hopkins, P. B. *Biochemistry* **1990**, *29*, 9225–9233.

43. Norman, D.; Live, D.; Sastry, M.; Lipman, R; Hingerty, B. E.; Tomasz, M.; Broyde, S.; Patel, D. J. *Biochemistry* **1990**, *29*, 2861–2875.

44. Culvenor, C. C.; Downing, D. T.; Edgar, J. A.; Jago, M. V. *Ann. N.Y. Acad. Sci.* **1969**, *163*, 837–847.

45. Mattocks, A. R. *J. Chem. Soc. C* **1969**, 1155–1162.

46. White, I. N. H.; Mattocks, A. R. *Biochem. J.* **1972**, *128*, 291–297.

47. Petry, T. W.; Bowden, G. T.; Huxtable, R. J.; Sipes, I. G. *Cancer Res.* **1984**, *44*, 1505–1509.

48. Borowy-Borowski, H.; Lipman, R.; Chowdary, D.; Tomasz, M. *Biochemistry* **1990**, *29*, 2992–2999.

49. Woo, J.; Sigurdsson, S. Th.; Hopkins, P. B. *J. Am. Chem. Soc.* **1993**, *115*, 3407–3415.

50. Tomasz, M. In *Advances in DNA Sequence Specific Agents,* Vol. 1; Hurley, L. H., Ed.; JAI Press: Greenwich, 1992, pp. 247–261.

51. Baker, B. F.; Dervan, P. B. *J. Am. Chem. Soc.* **1989**, *111*, 2700–2712.

52. Terano, H.; Takose, S.; Hosoda, J.; Kohsaka, M. *J. Antibiot.* **1989**, *42*, 145–148.

53. Shimomura, K.; Manda, T.; Mukumoto, S.; Masuda, K.; Nakamura, T.; Mizota, T.; Matsumoto, S.; Nishigaki, F.; Oku, T.; Mori, J.; Shibayama, F. *Cancer Res.* **1988**, *48*, 1166–1172.

54. Masuda, K.; Nakamura, T.; Shimomura, K.; Shibata, T.; Terano, H.; Kohsaka, M. *J. Antibiot.* **1988**, *41*, 1497–1499.

55. Fukuyama, T.; Goto, S. *Tetrahedron Lett.* **1989**, *30*, 6491–6494.

56. McClure, K. F.; Danishefsky, S. J. *J. Org. Chem.* **1991**, *56*, 850–853.

57. Williams, R. M.; Rajski, S. R. *Tetrahedron Lett.* **1992**, *33*, 2929–2932.

58. Williams, R. M.; Rajski, S. R. *Tetrahedron Lett.* **1993**, *34*, 7023–7026.

59. Woo, J.; Sigurdsson, S. Th.; Hopkins, P. B. *J. Am. Chem. Soc.* **1993**, *115*, 1199–1200.

60. Huang, H.; Pratum, T. K.; Hopkins, P. B. *J. Am. Chem. Soc.* **1994**, *116*, 2703–2709.

61. Huang, H.; Rajski, S. R.; Williams, R. M.; Hopkins, P. B. *Tetrahedron Lett.* **1995**, *35*, 9669.

INDEX

Printed and bound by CPI Group (UK) Ltd, Croydon, CR0 4YY

03/10/2024

01040436-0002